Fuels and Petrochemicals Division 2021

Held at the 2021 AIChE Annual Meeting

Boston, Massachusetts, USA and Online
7 - 11 November and 15 - 19 November 2021

ISBN: 978-1-7138-5710-5

Printed from e-media with permission by:

Curran Associates, Inc.
57 Morehouse Lane
Red Hook, NY 12571

Some format issues inherent in the e-media version may also appear in this print version.

Printed with permission by Curran Associates, Inc. (2022)

For permission requests, please contact AIChE
at the address below.

AIChE
120 Wall Street, FL 23
New York, NY 10005-4020

Phone: (800) 242-4363
Fax: (203) 775-5177

www.aiche.org

Additional copies of this publication are available from:

Curran Associates, Inc.
57 Morehouse Lane
Red Hook, NY 12571 USA
Phone: 845-758-0400
Fax: 845-758-2633
Email: curran@proceedings.com
Web: www.proceedings.com

TABLE OF CONTENTS

628859 Caustic Aqueous Phase Ethanol Reforming for Process-Intensified Hydrogen Production

Abstract:

Efficient generation of compressed pure hydrogen is a critical milestone for sustaining hydrogen in the energy network. Many of the traditional methods for compressed hydrogen production are hindered by high operating temperature, high energy cost, low efficiency, and many unit operations. The Caustic Aqueous Phase Ethanol Reforming (CAPER) process converts liquid-phase, oxygenated hydrocarbons to high pressure and high purity hydrogen at lower temperatures than traditional methods and using renewable resources. Additionally, the carbon dioxide produced by ethanol oxidation is captured within the caustic electrolyte solution. The electrolyte can be refreshed to release carbon dioxide or bicarbonate in a regenerator unit. Without the need of downstream separation, compression, or membranes, the CAPER process achieves process intensified hydrogen production at high efficiency. Shown in the figure below is the batch reactor for catalyst screening efforts and an overview of the electrochemical process. Ethanol is oxidized to carbon dioxide on the anode and hydrogen is formed at high pressure on the cathode. By operating at high pressure, the reactants remain in liquid phase while the only product gas is compressed hydrogen. The reduced operating potential provides many energy benefits over other electrochemical hydrogen production methods. The results of this work demonstrate a proof of concept of the CAPER technology and provide insights optimizing this system.

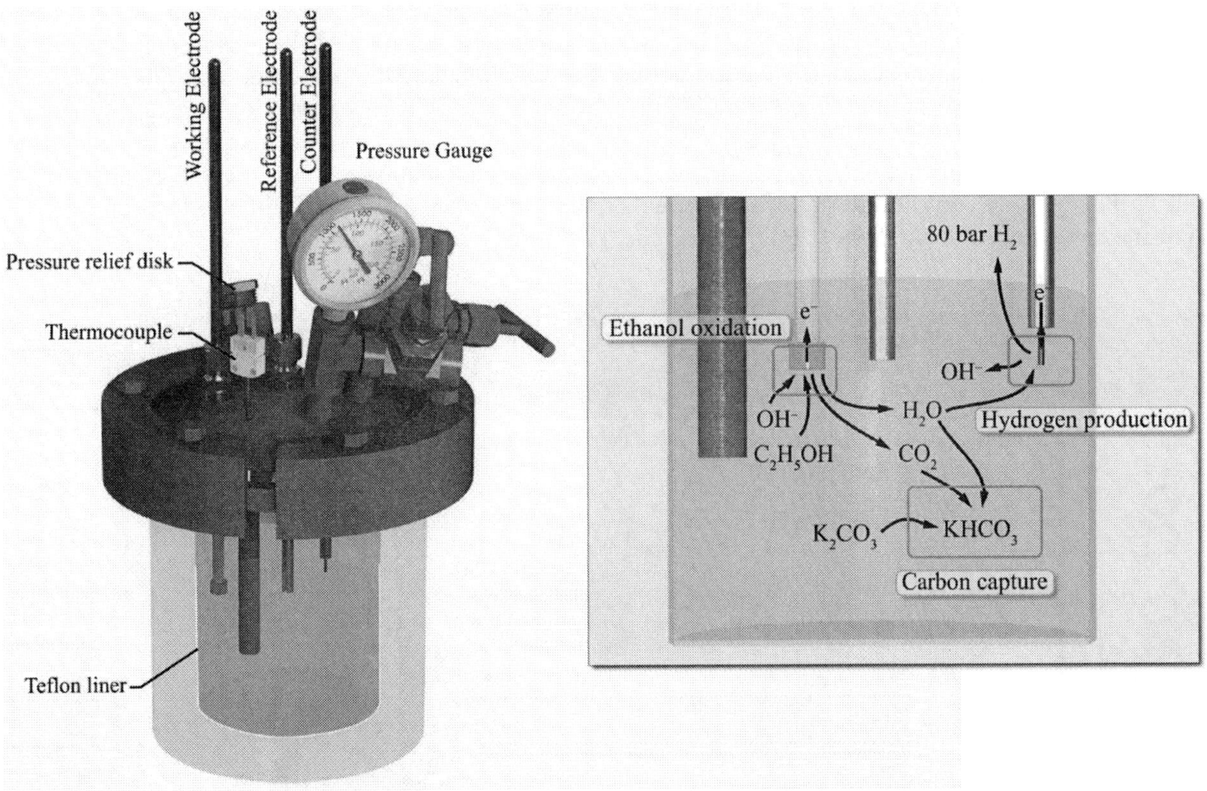

630863 Spatiotemporal Decoupling of Cerium-Mediated Water Electrolysis for Grid Energy Storage and Hydrogen Generation

Abstract:

The implementation of electrolysis systems for hydrogen production has continued to grow as the paradigm shift towards renewable energy and fuels progresses. However, when intermittent renewable energy sources power conventional polymer electrolyte membrane (PEM) electrolysis systems, the performance and safety of PEM electrolyzers degrade due to gas crossover. In order to make electrochemical hydrogen production safer when using intermittent renewable energy, decoupled electrolysis systems have been introduced to temporally and spatially separate the evolution of hydrogen and oxygen. This makes the electrolysis system safer, but it sacrifices efficiency compared to conventional electrolysis systems. In this presentation, I will discuss a cerium-mediated decoupled electrolysis system that can both produce hydrogen and store energy in the redox couple. Cerium was chosen due to its large redox potential of 1.6V, which is higher than water oxidation. This means that in one cell (charging cell), an energy input allows cerium to be oxidized while hydrogen is reduced, and in the other cell (discharging cell), energy can be generated by reducing cerium and oxidizing water. By combining experimental results and grid modeling, we identify optimal grid-integration strategies for the decoupled electrolysis system and provide guidelines for its dynamic operation with variable renewable electricity sources. Furthermore, since the oxidation of cerium and water are competing in the charging cell, Faradaic efficiency (FE) for the cerium reaction is less than 100%, which limits total efficiency of the system. To improve the FE of cerium oxidation, dynamic potential pulsing was employed to maintain a high concentration of cerium at the electrode surface improving the FE. The four variables that affect the behavior of the pulses include the oxidation pulse voltage, the rest voltage, the oxidation pulse duration, and the rest duration. Since the effect of these four factors on the FE and cerium conversion rate were unknown, we employed a Bayesian optimization (BO) to approach to identify the optimum combination of the four variables. The BO method was combined with electrochemical models of the reaction to accelerate the search. Finally, the optimal results will be put into context in terms of their effect on the performance and feasibility of the proposed cerium-mediated decoupled water electrolysis and energy storage system.

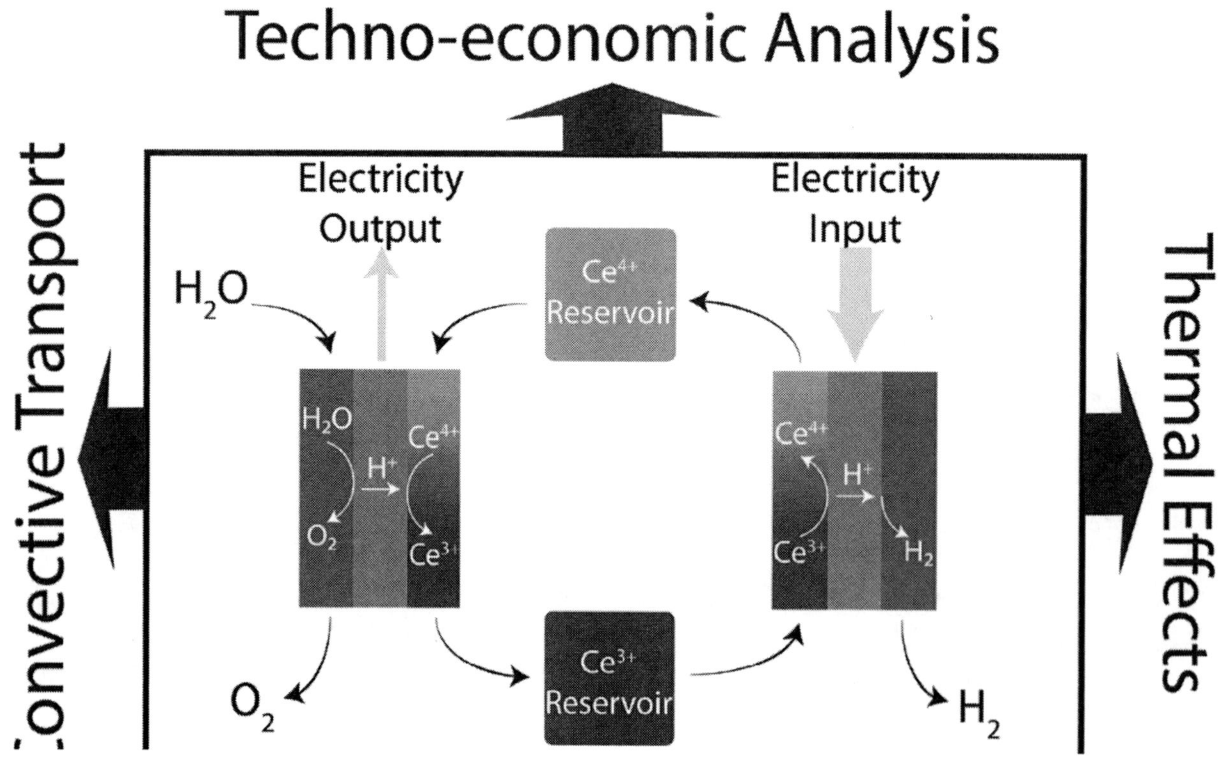

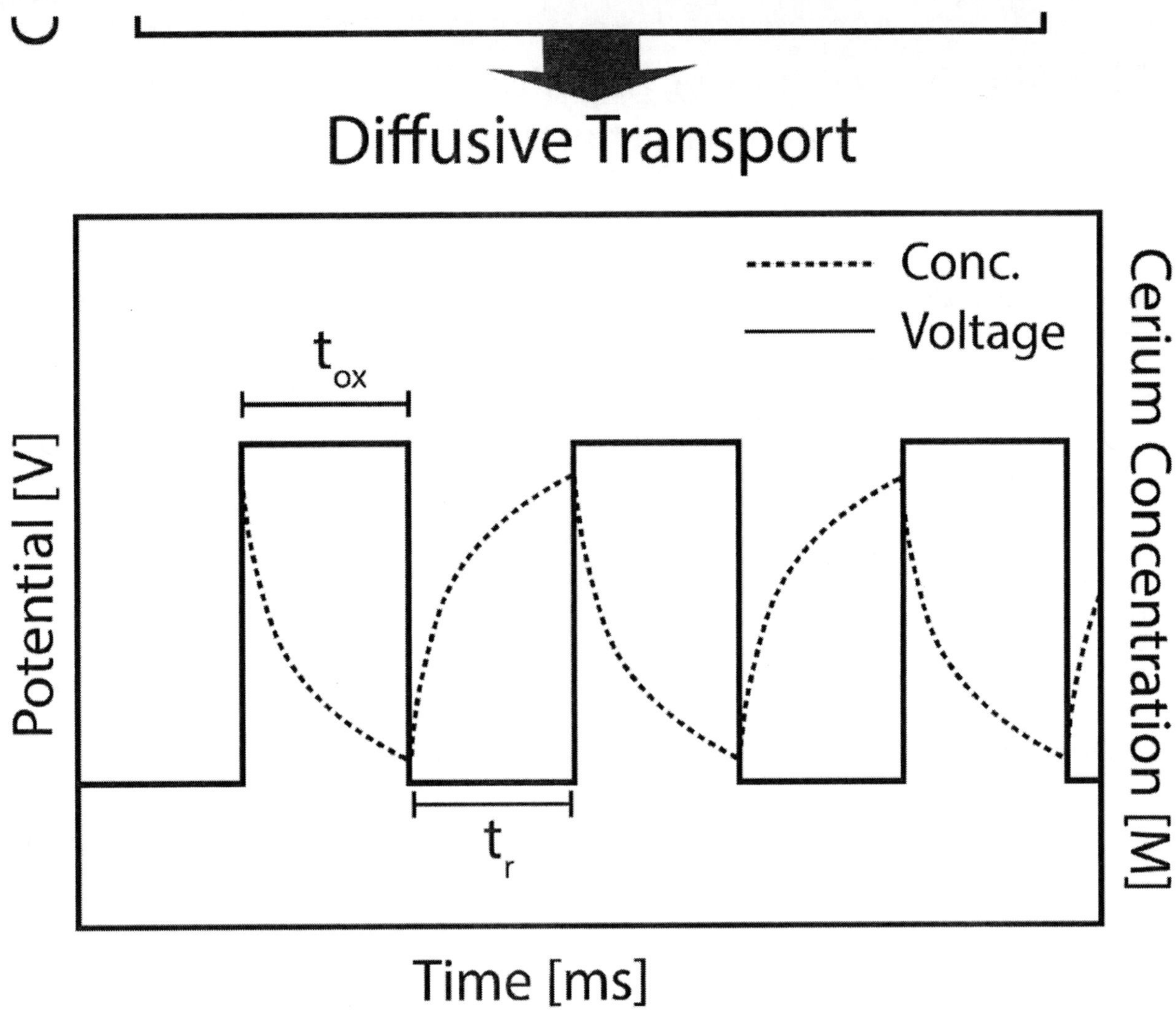
C
Diffusive Transport
Conc.
Voltage
t_{ox}
t_r
Potential [V]
Cerium Concentration [M]
Time [ms]

633153 Simulation Study of Chemical Transesterification Reaction for Biodiesel Production from Castor Oil

Abstract:

Biodiesel is a renewable source of energy and can be used directly without modification in engine. It is therefore an attractive alternative to the use of fossil fuels and gaining considerable attention. Biodiesel can be produced from a variety of feedstock through different transesterification systems. The production of biodiesel from castor oil through chemical based transesterification is expected to be a convenient route for many countries. However, the study on this transesterification reaction route is limited. The present study was therefore aimed to assess the several aspects of this transesterification reaction through simulation. In particular, the individual effects of reaction temperature, activation energy and Arrhenius constant on the composition and flow of biodiesel stream were analyzed and optimized. Alcohol (ethanol/methanol) and base (sodium hydroxide/ potassium hydroxide) were also altered to determine the impacts on biodiesel. The findings of this study would be useful for in-depth understanding of the reaction for the production of biodiesel from castor oil via chemical transesterification.

624879 Meeting U.S. Liquid Fuel Requirements Using Biomass Feedstocks to Replace Crude Oil in Large Integrated Refineries

Abstract:

We are developing a strategy [1, 2] to replace crude oil feedstocks with biomass feedstocks in large integrated refineries to achieve two goals. The first goal is to use low-carbon biofuels to economically replace all liquid hydrocarbon fuels (gasoline, diesel, marine fuel and jet fuel). The second goal is to provide a market-based method for negative carbon emissions; that is, carbon dioxide is removed from the atmosphere and sequestered based on market prices for negative carbon emissions.

These goals require large-scale nuclear biorefineries (250,000 barrel per day equivalent) that can economically produce a variable product slate (hydrocarbon fuels, hydrocarbon chemical feedstocks and sequestered carbon dioxide) based on market prices. If biomass feedstock costs or liquid fuel prices are low in a particular year, more pure carbon dioxide will be produced for sequestration assuming there is a market for negative emissions. If hydrocarbon fuel prices are high, there may be no carbon dioxide sequestered in a particular year. The market (including price of sequestered carbon dioxide) determines the relative quantities of different products. The core of the refinery is similar or identical to existing large integrated oil refineries with two major changes.

Front end processing. The front end converts biomass into a synthetic crude oil. The baseline process is Fischer Tropsch (gasifier, water shift reactor and FT synthesis) to produce the hydrocarbon "crude oil" for the refinery. The Fischer Tropsch process is used today to convert coal and natural gas into liquid hydrocarbon fuels at a commercial scale. Varying heat and hydrogen input to the gasifier and water-shift reactor shifts relative production of hydrocarbons and carbon dioxide. There are alternative processes to convert different types of biomass into synthetic crude oil for the refinery.

Nuclear heat and hydrogen. Nuclear energy is used to provide heat and hydrogen for all refinery operations. Nuclear heat and hydrogen input at the refinery is the enabling technology for sufficient biomass feedstocks to meet liquid fuel demands without major impacts on agriculture and forestry, including food and fiber costs and availability. This objective is achieved by two mechanisms. Biomass has a typical composition near $CH_{1.44}O_{0.66}$ that must be converted to a hydrocarbon fuel with a composition near CH_2. The oxygen can be removed by using carbon in the biomass to create carbon dioxide. We can remove more carbon to obtain the proper carbon to hydrogen ratio and provide energy for the process. Alternatively we can add external heat and hydrogen to operate the process and remove the oxygen as water. This requires massive amounts of heat and hydrogen—but this approach can double liquid hydrocarbon fuels per unit of biomass feedstock. The only low-carbon sources of concentrated heat and hydrogen at the required scale are nuclear and fossil fuels with carbon capture and storage (CCS). The second consideration is that there are many sources of biomass that are good carbon sources but relatively poor energy sources (double-cropped biomass, sewage sludge, garbage, etc.). These are viable low-carbon feedstocks if external heat and hydrogen inputs to the biorefinery are provided by nuclear plants.

The required large-scale nuclear biorefineries require viable systems to deliver biomass to the refinery front gate [Fig. 1]. Large nuclear-powered biorefineries require massive quantities of feedstocks, on the order of 30,000 tons per day. There is insufficient low-density biomass within an economic transport distance to meet the feedstock demand of such large biorefineries. It is prohibitively expensive to ship low-density biomass long distances to such large refineries. What is required are intermediate processing facilities called "depots" to consolidate near the farm or forest the local biomass into stable high-density forms that can be economically shipped long distances to large-scale nuclear bio-refineries. This is equivalent to shipping dense storable #2 corn to ethanol plants. As a point of comparison, the largest corn milling plant receives about 15,000 tons of corn per day by unit train.

There are multiple potential depot processes to create a storable high-density intermediate product: pelletization, pyrolysis, and anaerobic digestion—each sending different intermediate high-density intermediate products to the nuclear bio-refinery with different secondary products. These secondary "wastes" include biochar and anaerobic digester waste that are fertilizers to recycle stable carbon, phosphorus and other nutrients back to the land. Biochar can be used to build up soil and sequester carbon or as a feedstock for the refinery—depending upon market incentives for negative carbon emissions.

Depot processes are directly coupled to agricultural processes to assure long-term sustainability. The storable shippable product creates a biomass refinery feedstock market similar to that for corn or, on much smaller scale, wood chips. The system fits within the existing American agricultural system and creates a new market for agriculture. Since the 1930s, the primary challenge of American agriculture has been excess production capacity. The excess production capacity now has a very large potential market—providing energy services and low carbon feedstocks for petrochemical production.

The full potential of this system to produce biomass is not well understood. Historically, biomass studies have examined the availability of biomass as an energy source. Such studies indicate that biomass could provide up to a quarter of global energy demand. The question here is different: What is the potential to provide enough carbon from biomass to replace all liquid fuels, hydrocarbon feedstocks to the chemical industry and negative carbon emissions—where we have massive hydrogen and heat input from nuclear sources to convert that biomass into useful products? Total global biomass produced annually by photosynthesis is about 100 billion tons; thus, the total resource base is sufficient. The question is price. This question has two subcomponents. What could American agriculture produce if there was a market? What is the carbon biomass from feedstocks that are not good fuels but excellent carbon sources? Preliminary studies indicate sufficient feedstocks to meet world liquid fuel demand—provided nuclear heat and hydrogen inputs at the refinery so the biomass is first considered as a carbon feedstock and secondarily as an energy source. The at-the-farm or forest biomass prices per ton today are similar to crude oil prices.

We have started to undertake studies and workshops to map out the system options with base-line and alternative options—and the limits to each option. Where are the bottlenecks? What are the primary economic drivers? Some parts of this system are well-understood; eg, large-scale industrial technologies such as integrated refineries. Some parts of this system such as Fisher Tropsch are existing technologies used in some markets but not for this specific market at this scale. Other components such as depot processes have been demonstrated at pilot scale.

The other requirement is to understand the economics and market drivers. This includes the potentially future market for negative carbon emissions. The expectation is that ultimately there will be a tax on carbon dioxide emissions. If there is a tax, there should logically be a market for negative carbon emissions; that is, removing carbon dioxide from the atmosphere and sequestering carbon dioxide. The economics of carbon capture with fossil fuel power plants is that carbon capture is expensive (carbon dioxide diluted with air) but carbon sequestration is relatively cheap if a reasonably pure stream of carbon dioxide is sequestered. Refinery biomass gasifiers and the water gas shift reaction to provide syngas (a mixture of hydrogen and carbon dioxide) to the Fischer Tropsch process can be operated to provide various ratios of syngas and carbon dioxide—based on heat and hydrogen inputs. If there is a large market for negative emissions, it provides the refinery two revenue streams—fuels and negative carbon emissions where variable output is based on biomass feedstock costs and the prices for the two products. Negative carbon emissions become the variable product that in some years would be a major product but in other years perhaps little or no production. From a climate perspective, what is important is negative emissions over decades—not over a single year.

[1] C. Forsberg and B. Dale, "Replacing Liquid Fossil Fuels with Liquid Biofuels from Large-Scale Nuclear Biorefineries" *Volume 8; Proceedings of Applied Energy Symposium: MIT A+B, United States 2020*, August 12-14, 2020, Cambridge, MA. https://applied-energy.org/mitab2020/program

[2]. C. Forsberg et al., *Workshop: Can a Nuclear Biofuels System Enable Liquid Biofuels as the Economic Low-carbon Replacement for all Liquid Fossil Fuels and Chemical Industry Inputs*, Webinar (August 4, 11, 18), Massachusetts Institute of Technology, Michigan State University, North Carolina State University. (Proceedings in preparation)

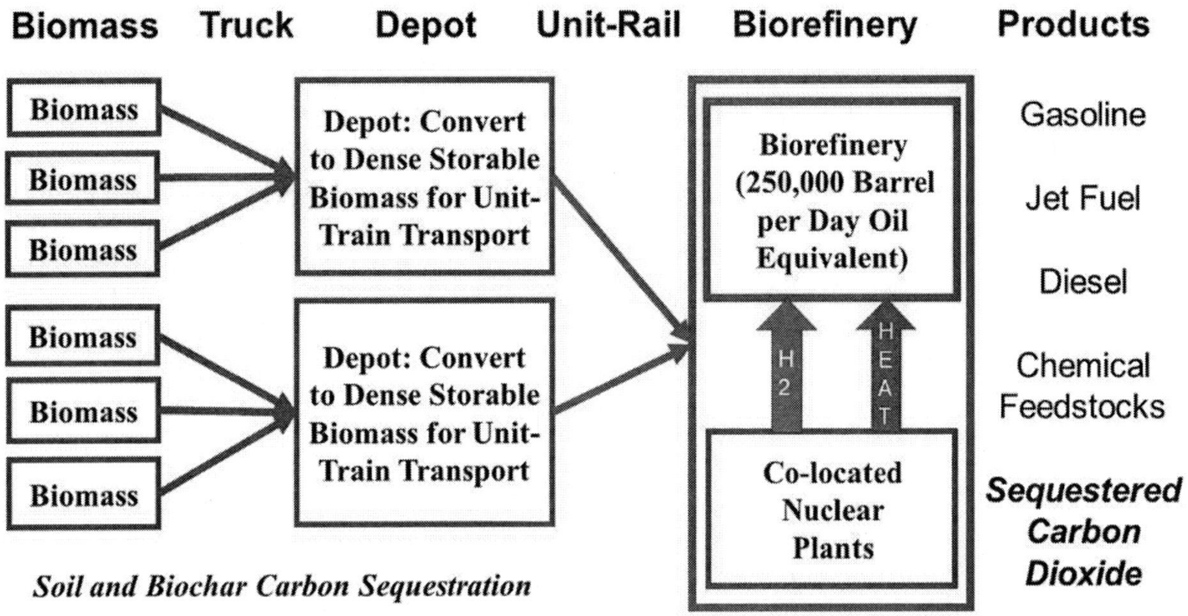

Fig. 1. Nuclear Biomass Refinery System

628754 Synthesis of Renewable Jet Fuel Range Cycloparaffins with Levulinic and Formic Acids

Abstract:

This research presents a production pathway for renewable cycloparaffinic fuel blendstocks for military and commercial jet fuels. Cycloparaffins are major constituents of jet fuel that provide beneficial properties to the fuel such as low freezing point, increased density, higher volumetric heating value, and enhanced operability. Both aromatics and cycloparaffins are known for contributing to elastomer swell which is an important fuel property required for the proper operation of aircraft fueling systems. Most efforts to produce bio-jet fuels have focused on producing straight-chain and branched-chain alkanes because of their higher energy densities. However, there is still a need for technologies to produce renewable cycloparaffins for sustainable full-performance military and commercial jet fuels.

In this work, jet fuel range cycloparaffins were synthesized with levulinic and formic acids which can be produced from lignocellulosic biomass. Using a novel non-catalytic thermochemical process a mixture of levulinate and formate salts was converted to synthetic crude Thermal DeOxygenation (TDO) oil in a pilot-scale continuous flow reactor at 450 °C and atmospheric pressure. The crude TDO oil has a broad boiling point range (75-550 °C), contains a very low oxygen content (< 5 wt %) and has high levels of monoaromatics and polyaromatics. Crude TDO oil was hydroprocessed using a commercial Ni/SiO_2-Al_2O_3 catalyst in a bench-scale continuous flow reactor. The hydrogenation reactions were carried out at 300 °C and 750 psig with an excess flow of hydrogen at 300 sccm. The feed flow rate was maintained at 0.3-0.5 mL/min with a corresponding weight hourly space velocity (WHSV) of 0.1 - 0.3 hr^{-1}. Hydrogenated TDO oil was distilled to produce jet fuel range (180-250 °C) hydrocarbons according to the American Society for Testing and Materials (ASTM) standards.

The jet fuel fraction (180-250 °C) of hydrotreated TDO oil has a 32.5 % mass recovery and mainly consists of 94 vol% cycloparaffins. The remainder of the jet fuel fraction consists of 5.73 vol% aromatics with trace n-paraffins and iso-paraffins. The cycloparaffins range from C_9-C_{15} and consist of a mixture of monocycloparaffins (10 vol%), dicycloparaffins (78.47 vol%) and tricycloparaffins (5.52 vol%). The jet fuel range cycloparaffins have a very low freezing point (-70 °C) and low viscosity (3.9 mm^2/s). The TDO oil jet fuel fraction is compliant with all ASTM 3699 specifications and meets MIL-DTL-83133H specifications with the exceptions of density, hydrogen content, and higher heating value. Based on the fuel properties the TDO oil cycloparaffinic fuel can be blended up to 50 % by volume with JP-8 as well as Jet-A and meet the specific fuel standards.

620511 Influence of Alcohol Additives on the Viscosity and Solubility of Ethanol/Diesel Fuel Blends: A Molecular Dynamics Simulation Study

Abstract:

The viscosity and solubility of ethanol/diesel fuel blends have an important influence on its application in diesel engines. Molecular dynamics simulation was employed to study the effects of different alcohol additives on the viscosity and solubility of the blends between diesel and anhydrous ethanol. Among these alcohols, n-butanol, n-heptanol and n-decanol were selected for investigating the influence of alcohol chain length on solubility and viscosity, while the effects of straight and branched chain on solubility were studied via comparing with four isomers of butanol.The results showed that the viscosity of the blends increased with the increasing of alcohol carbon number. The analysis of diffusion coefficient represented that the alcohol additives enhanced the molecular mobility and activity, and alcohols with higher carbon numbers provided a better inter-soluble capacity. It was clearly that the straight chain butanol had better miscibility than that of branched structures. The analysis of energy revealed that the hydrogen bonding interaction and Van der Walls attracting had significant effects on enhancing miscibility of blends. The present study provides fundamental understanding of the interactions for ethanol/diesel blends with co-solvents, revealing the mechanism of difference in solubility and micro behavior of diesel molecules in fuel blends.

626554 Microwave-Assisted Demulsification of Tight Crude Oil-Water Emulsions

Abstract:

The main objective of this study is to use microwaves to develop an energy efficient and economical method to demulsify stable crude oil and water emulsions. These crude oil emulsions are formed by high shear mixing of process water and crude oil during pumping, desalting and transportation through wellbores and pipelines and are stabilized by emulsifying agents that form interfacial films at the oil-water interface. These surface-active species, such as asphaltenes, resins, solids, waxes, or commercially available chemical emulsifiers present in the crude, are responsible for the formation of stable emulsions. Tight crude oil emulsions, characterized by suspended droplets on the order of submicron to tens of microns, often cause serious problems to the wellbores and pipelines such as clogging, corrosion, and pump failures. To simulate field conditions, we mixed two crude oils to prepare samples with American Petroleum Institute (API) gravities between 28.9° and 46.0°. These crudes were mixed with water to create emulsions containing between 20% and 80% water. This array of Water-in-Oil (W/O) and Oil-in-Water (O/W) emulsions with different densities and viscosities were mixed at varying shear rates. A series of batch demulsification runs of these crude oil emulsions were done by exposure to microwaves under a well-defined set of parameters such as temperature, power, time, and salinity. To determine the efficiency of phase separation after treatment, final oil-phase water content was measured using Karl Fischer Titration and the water-phase oil content was determined using an Oil and Grease analyzer. Under similar hold times and temperatures, microwave separation methods more effectively separated the phases than thermal separation methods. Energy input analysis was performed from the Power-Time data obtained from each demulsification run. A separation efficiency model was developed based on the energy and mass balance study for the microwave demulsification method.

624867 Optimal Control for Pipeline Flushing Operations in Lubricants Blending and Packaging Industries

Abstract:

Interfacial mixing of consecutive product batches in multiproduct pipeline systems has been a long-standing and economically significant issue in the petroleum industries (Li and Bidmus 2019; He et al. 2019). The U.S. liquid petroleum pipeline comprises 200,000 miles of pipe in all fifty states. These pipelines are vital to the nation's economy and transport a wide range of materials, including gasoline, aviation fuels, kerosene, diesel fuel, and heating oil (Pharris and Kolpa 2008). In this work, we investigate one such multiproduct complex network of pipelines used to transport multiple fluids. Ideally, the production lines must be cleaned/flushed between product changeovers to avoid contamination and maintain final product integrity. The flushing operation utilizes a significant amount of finished products, which results in the contamination of the high-value products, and waste in the production operation. The process results in high economic losses to the industries and adds to a significant environmental burden. Therefore, it is crucial to understand the drawbacks of the existing flushing operation and minimize the amount of oil that gets downgraded. Hence, our research goal is to implement novel process optimization techniques and enhance the existing flushing operations in lubricant industries by formulating the operation as an optimal control problem.

The current flushing operation is based on previous operator experience to determine appropriate flush times and volumes. This, however, is not optimal, and there are potential opportunities for cost, energy, and material savings on detailed analysis and subsequent optimization of this process. In this research, we are developing models of fluid hydrodynamics in multiproduct pipelines. We will use process optimization tools to predict strategies for conducting the flushing operations with less energy consumption and material downgrade.

This research aims to develop an optimal control problem for flushing operations in the petroleum industries. The problem is formulated using various performance indices: minimum flush time, maximum product purity, and minimum oil downgrade. We use Pontryagin's maximum principle for solving these problems. In our work, we present the formulation of these optimal control problems and analysis of the solutions. Our objectives can be categorized as follows:

- Data collection and analysis
- Model development for fluid hydrodynamics within the flushing process
- Formulation of an optimal control problem that minimizes the flushing time, maximizes the product purity, and minimizes the oil downgrade
- Solution strategy using Pontryagin's maximum principle and
- Solution analysis and recommendations for industrial application

A 23-week lube oil flush study data was collected from our industrial partner. This data was analyzed to understand the existing flushing operations at the facility. The uncertainty observed through the data analysis gave us a strong indication that the current flushing processes can be optimized, thus minimizing the oil downgrade and overall energy consumption. Following this analysis, on-site flushing experiments were designed, and 70 product changeovers were conducted to identify the key process conditions and bottlenecks. We then developed a first-principles model of the flushing operation, which incorporated the essential parameters from the experimental flush study. The parameters include but are not limited to the viscosity of individual products, the viscosity of product blend, density, flowrate, flushing time, and downgraded/flushed oil. These parameters are incorporated into time-dependent differential equation model, and optimal control problems are formulated using various performance indices mentioned above. Pontryagin's maximum principle is used for solving these optimal control problems (Diwekar 2008; Yenkie and Diwekar 2013). The advantage of this method over other approaches of solving optimal control problems is that it does not involve second-order differential equations or partial differential equations

(Benavides and Diwekar 2013; Yenkie and Diwekar 2014).

Our research can be an excellent starting point in optimizing the existing flushing operations in the lubricant industries. It can lead to financial savings for the company and improve the environmental footprint of the process.

Keywords: Flushing, lubricants, optimal control, optimization

Acknowledgments

- S. Environmental Protection Agency's Pollution Prevention (P2) Program
- Sustainable Design and Systems Medicine Lab
- Undergraduate Students: Joseph D'Intino, Anthony Wiley, Jacob Martin, Erik Dunn, Emily Rooney, Diana Castro, Marissa Martine, Spencer Verdoni
- Rowan University Department of Chemical Engineering

References

Pahola B. T., and Diwekar U. 2013. "Studying Various Optimal Control Problems in Biodiesel Production in a Batch Reactor under Uncertainty." *Fuel* 103 (January): 585–92. https://doi.org/10.1016/j.fuel.2012.06.089.

Diwekar U. 2008. *Introduction to Applied Optimization.* Springer.

He G., Na Y., Kexi L. Baoying W., and Liying S. 2019. "A Novel Numerical Model for Simulating the Quantity of Tailing Oil in the Mixed Segment between Two Batches in Product Pipelines." Research Article. Mathematical Problems in Engineering. Hindawi. August 19, 2019. https://doi.org/10.1155/2019/6892915.

Li G., and Hamid B. 2019. "Estimating Mixing Lengths in Multi-Product Pipelines." In . OnePetro. https://onepetro.org/PSIGAM/proceedings/PSIG19/All-PSIG19/PSIG-1903/2103.

Pharris T. C., and Kolpa. R. L.2008. "Overview of the Design, Construction, and Operation of Interstate Liquid Petroleum Pipelines." ANL/EVS/TM/08-1, 925387. Argonne National Lab. https://doi.org/10.2172/925387.

Yenkie, K. M., and Diwekar U. 2013. "Stochastic Optimal Control of Seeded Batch Crystallizer Applying the Ito Process." *Industrial & Engineering Chemistry Research* 52 (1): 108–22.

Yenkie, Kirti M., and Diwekar U. 2014. "Optimal Control for Predicting Customized Drug Dosage for Superovulation Stage of in Vitro Fertilization." *Journal of Theoretical Biology* 355 (August): 219–28. https://doi.org/10.1016/j.jtbi.2014.04.013.

Graphical Abstract

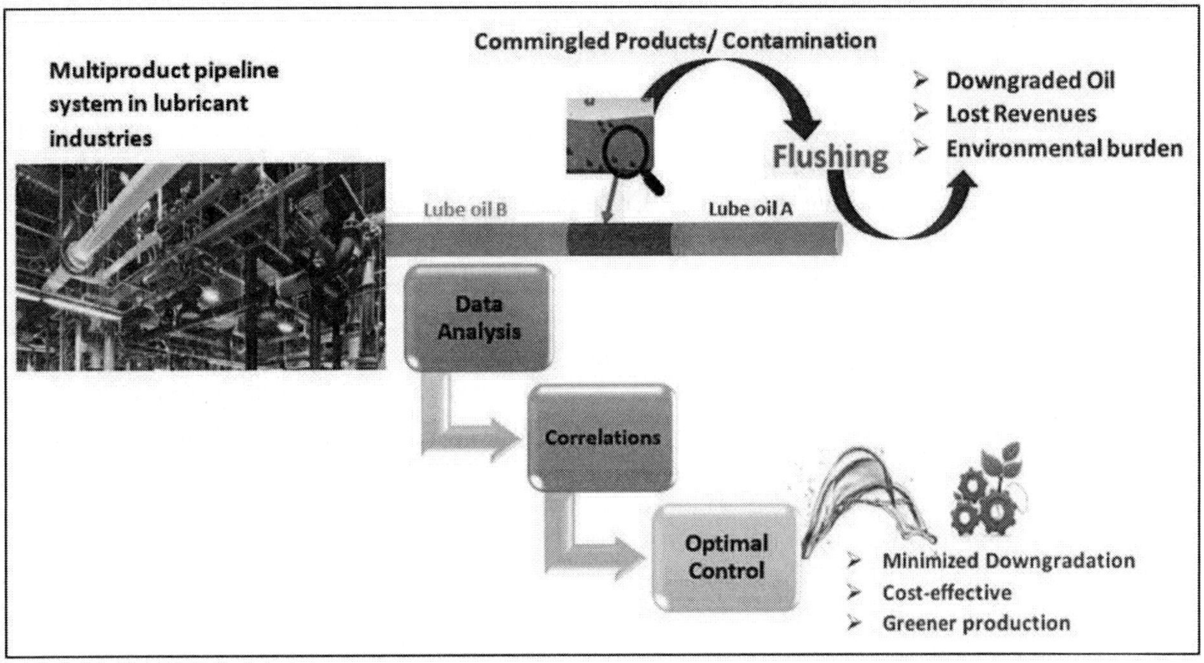

CFD and Experimental investigation on Gas-Oil-Water three-phases flow toward the onset of liquid film reversal in a vertical pipe

Chuan Xie [a,b,1,*], *Yonghui Liu* [a,2,*], *Ning Wu* [a,3], *Jianying Yang* [c,4], *Man Chen* [c,5], *Zhiguo He* [c,6], *Fanhua Zeng* [b,7],

[a] State Key Laboratory of Oil & Gas Reservoir Geology and Exploitation, Southwest Petroleum University, Chengdu, Sichuan 610500, China

[b] Petroleum Systems Engineering, Faculty of Engineering and Applied Science, University of Regina, Regina, Saskatchewan S4S 0A2, Canada

[c] Sichuan Changning Natural Gas Development Co, Ltd., Chengdu 610051, P.R. China

Abstract

Liquid loading is a common problem for condensate gas wells as natural gas is always produced with water and condensate. Accurate prediction of liquid loading and proper deliquification measurements are of great significance to stabilize the gas production for gas wells in condensate gas reservoirs. In this paper, Eulerian-Eulerian Multi-Fluid VOF approach was proposed to predict the critical velocity of film reversal, which refer the onset of liquid loading. Experimental tests were also conducted to provide data for flow pattern and pressure gradient measurements to validate the numerical simulation model. All tests were conducted based on the visual pipe with a height of 5 m and an inner diameter of 50 mm. The experimental results show that the oil-water-gas flow pattern and pressure gradient agree well with the simulation results. CFD has been carried out to simulate critical velocity of film reversal under the annular flow condition in a 2D vertical geometric model of 62 mm in diameter with different water and oil superficial velocities. The velocity profiles show that the slip between the water phase and oil phase can be neglected near the walls, but there is a significant slip phenomenon between gas phase and liquid phase in this region. The simulation results show that the value of surface tension and drag coefficient affect the accuracy of the critical velocity of film reversal. For this reason, the Schiller and Naumann drag force model was applied in simulation method to describe the drag law among condensate oil-water-gas three phases respectively. Compared with the public data of critical velocity in the literature, the absolutely error of simulations was below 15% without considering emulsification of oil-water.

Key words: Liquid loading; Condensate gas reservoir; CFD (Computational Fluid Dynamics); Critical velocity

Introduction

In condensate gas reservoirs, with the development of gas wells coming into middle and late stage, the reservoir pressure decreased, condensate oil and water flow into wellbore. In this way, the wellbore experience oil-water-gas three phase flow, with the gas flow rate decreasing, the liquid cannot be carried to the surface because of the low gas velocity. Meanwhile, the intense pressure fluctuant occurs in tube, which means the gas well is suffering from liquid loading. Furthermore, with the liquid accumulate at the bottom of wellbore, resulting in gas well have to shut down [1]. Therefore, it is important to predict the onset of liquid loading to maintain the stable production of gas wells.

To predict the liquid loading onset value, two main types of models were proposed. One of that is based on the force balance analysis of droplet in the gas core. The most basically semi-empirical model was conducted by Turner et al [2], according to set the critical Weber number as a value of 44 and drag coefficient value is 0.44. The modified model established by Coleman et al [3] to agree better with low pressure gas well, suggested that the 20% upward adjustment of Turner et al. (1969) model was unnecessary. Besides, more modified droplet models were proposed which based on the factors of drag coefficient, kinetic energy, flow pattern, weber number and the sharp of droplet [4-7]). Owing to the simple expression of droplet model, it is widely used by engineers to predict the onset of liquid loading in gas field. However, according to observation of experiments and gas well production, it is hard to prove the relationship of droplet reversal and liquid loading.

Another type of models was based on the liquid film reversal. Experimental investigations of annular/churn air–water upward flow was conducted by Van't Wenstende [8], the diameters of droplets were measured to compare with which assumed in Turner's model. The results shown that the size of measured droplet is much smaller than results of prediction, and there is no droplets reversal are observed in their experiments. Afterwards, the droplet entrainment rate was measured by Sawant et al. [9] and Alamu [10]. In their studies, they found droplet entrainment rate is much lower than liquid film under the annular condition. Form the conclusion of above researches, the film reversal should be explaining the phenomenon of liquid loading more reasonable.

Recently, a few scholars were studied the liquid loading by CFD mehod. The most evaluable research was conducted by Vieiro [11] with an axisymmetric 2D geometry model, and the liquid loading phenomenon was simulated an the ANSYS CFX version 13.0 platform with homogeneous model. In this study, they don't motion any clear criterions to define the onset of liquid loading. However, the numerical pressure gradient data agrees well with the experimental results of Westende [12].

However, when these models predict the critical velocity of condensate gas wells, because of they didn't take the condensate oil phase into consideration, the calculation results may cause unacceptable error. It is necessary to conduct a method to predict the onset of liquid loading under the oil-gas-water three phases condition. In this work, Eulerian-Eulerian Multi-Fluid VOF approach was proposed to predict the critical velocity of film reversal, which refer the onset of liquid loading. Experimental tests were also conducted to provide data for flow pattern and pressure gradient measurements to validate the numerical simulation model.

Problem description

In this paper, the inside of tube experiences the oil-gas-water three annular flow, and the main work is to predict the onset of film reversal which is the key criterion of liquid loading. As shown in Fig. 1 a, the liquid film attaches on the wall of tube and the condensate oil is entrain in water. The liquid droplet entrain in gas core and gas bubble in liquid film can be ignore in this case. The schematic diagram of the flow domain shown in Fig. 1 b, an axisymmetric 2D vertical geometry with oil, water and gas inlet separately. The width of the liquid inlet is set to 20mm to avoid jet flow at the inlet.

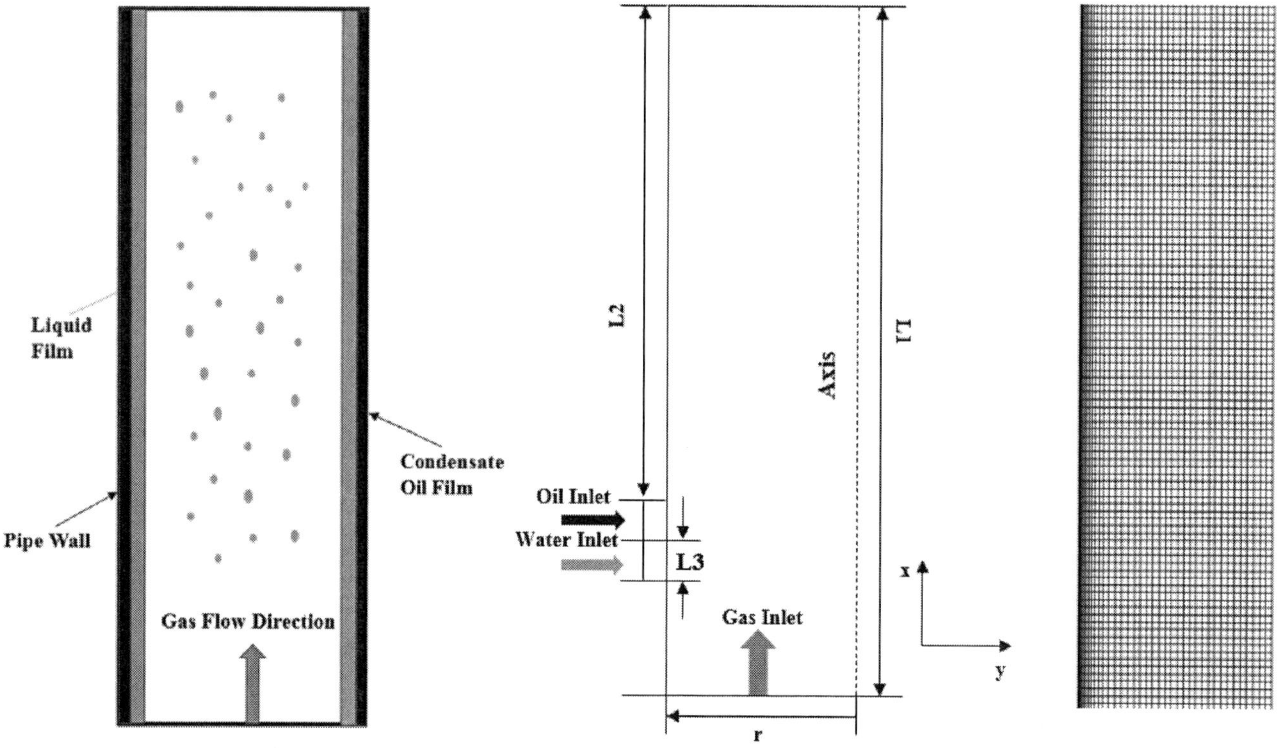

a. Flow regime b. Flow domain c. Mesh

Fig. 1. Schematic diagram of the flow domain

The computational domain is a vertical pipe of radius, r, and length, L. The domain is bounded by the pipe axis. All the parameters are presented in Table 2, including the gas and liquid properties.

Table 1

The computing field parameters and fluid physical properties

Parameter	Unit	Definition	Value
r	mm	Pipe inner radius	31
L_2	mm	Length of observation	1500
L_3	mm	Length of liquid inlet	20
ρ_g	kg/m^3	Gas phase density	1.225
μ_g	kg/ms	Gas phase kinematic viscosity	1.7894×10^{-5}
ρ_w	kg/m^3	Water density	992
μ_w	kg/ms	Water dynamic viscosity	1.03×10^{-3}
ρ_o	kg/m^3	Oil density	889
μ_o	kg/ms	Oil viscosity	1.06

The critical superficial gas velocity is defined as the corresponding U_{SG} to onset of film reversal in this study. For vertical pipes, since the drag force of the liquid film flow is provided by the gas core, the flow of the liquid film will show a gradually increasing velocity profile from the pipe wall to the inner interface. The diagram of the velocity profile of the liquid film under different gas flow rates is shown in Figure 2. Under the high gas flow rate condition, the gas core can provide a very high internal shear stress, resulting the gas-liquid interface has a strong momentum exchange. As shown in Figure 2 (a), the liquid film flows upward and the frictional resistance of liquid film along the pipe wall shown a downward trend. With the decreasing of gas flow rate, the drag force of gas core provides for the liquid film also decreased, which will cause the frictional resistance of the liquid film on the pipe wall decrease to zero. At this time, the gravity of the liquid film is equal to the drag force of the air core, as shown in Figure 2 (b). As the gas flow rate decreases further, the liquid film becomes thicker and more unstable, and the pipe wall provide upward frictional support for the liquid film. Therefore, when the the frictional resistance of the liquid film on the pipe wall is zero, the drag force of gas core is equal to the gravity, that is the boundary condition for solving the liquid film reversal.

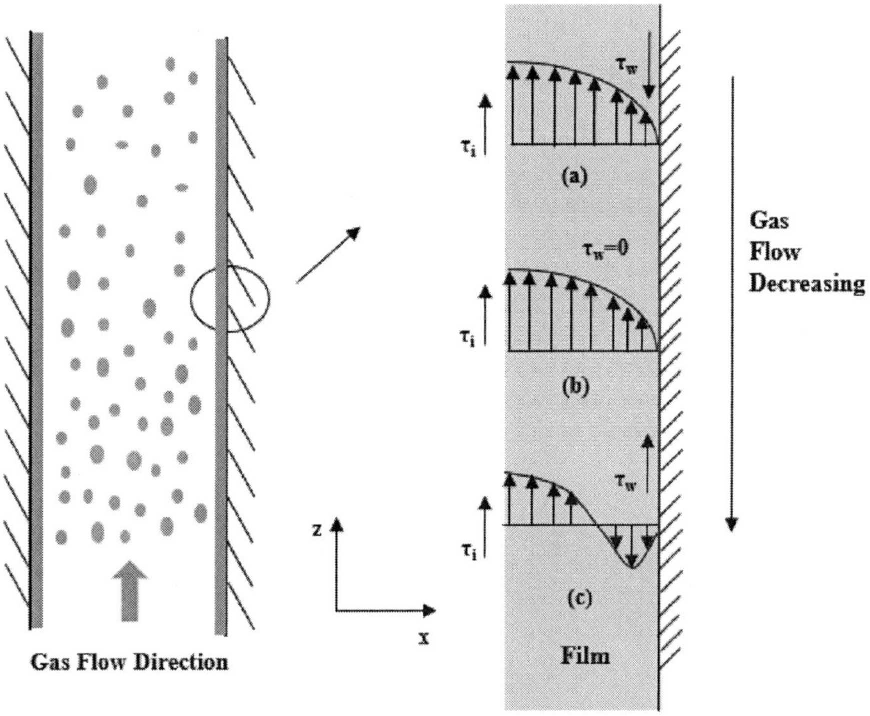

Fig. 2. Schematic diagram of the liquid film reversal

Mathematic model and Numerical method

Mathematic model

The flow process is assumed to be adiabatic and incompressible, and no mass transfer exists. Therefore, the governing equations of gas and liquid should be determined using mass and momentum conservation, respectively.

Mass conservation

$$V_q = \int_V \alpha_q dV \quad , (1)$$

$$\hat{\rho}_q = \alpha_q \rho_q$$

$$\frac{\partial}{\partial t}\left(\alpha_q \rho_q\right) + \nabla \cdot \left(\alpha_q \rho_q \vec{\upsilon_q}\right) = \sum_{p=1}^{n} \dot{m}_{pq}$$

Momentum conservation

$$\frac{\partial}{\partial t}(\alpha_q \rho_q \vec{v}_q) + \nabla \cdot (\alpha_q \rho_q \vec{v}_q \vec{v}_q)$$

$$= -\alpha_q \nabla p + \nabla \cdot \overline{\overline{\tau}}_q + \sum_{p=1}^{n} (\vec{R}_{pq} + \dot{m}_{pq} \vec{v}_{pq}) + \alpha_q \rho_q (\vec{F}_q + \vec{F}_{lift,q} + \vec{F}_{Vm,q}) \quad , (2)$$

Where τ_k is liquid phase stress–strain tensor expressed as

$$\overline{\overline{\tau}}_k = \alpha_k u_k \left(\nabla U_k + \nabla U_k^T\right) + \alpha_k \left(\lambda_k - \frac{2}{3}\mu_k\right) \nabla \cdot U_k \overline{\overline{I}} \quad , (3)$$

In this section, F_{CSF} refers to the surface force, and R_{GL} is the interaction between gas and liquid, which can be described as below

$$R_{GL} = \mathrm{K}_{GL}\left(U_G - U_L\right) \quad , (4)$$

$$K_{GL} = \frac{\alpha_L \alpha_G \rho_G}{\tau_G} f \quad , (5)$$

where:

$$\tau_G = \frac{\rho_G d_G^2}{18\mu_L} \quad , (6)$$

and

$$f = \frac{C_D Re_R}{24} \quad , (7)$$

Schiller and Naumann's [13] model were used to calculate the C_D (drag coefficient), which can be presented as

$$C_D = \begin{cases} \dfrac{24\left(1 + 0.15 Re^{0.687}\right)}{Re} & Re \le 1000 \\ 0.44 & Re \le 1000 \end{cases} \quad , (8)$$

$$\mathrm{Re}_R = \frac{\rho_{LG}\left|U_G - U_L\right| d_{LG}}{\mu_{LG}} \quad , (9)$$

Brackbill [14] model was used to solve the F_{CSF} (continuum surface force), which can be presented as

$$F_{CSF}=\sigma\frac{\rho_{LG}\kappa_L\nabla\alpha_L}{\frac{1}{2}(\rho_L+)\rho_G}, (10)$$

Numerical Method

As we all know, the Volume of fluid (VOF) model often apply to track the surface of each phase, but the interaction of multiphase is hard to calculate completely. The Eulerian method not only calculate the N-S equations for each phase separately, but also shared pressure, interphase momentum exchange and energy exchange. However, the Eulerian method cannot capture the interface of each phase accurately. ANSYS Fluent® 19.0 conducted a new approach to solve the multiphase flow problem, the Eulerian-Eulerian Multi-Fluid VOF overcomes some of the flaws of the VOF scheme as a result of the shared velocity and temperature formulations [15,16], hence, its adoption for the current study. As for the turbulence model, High turbulence usually exist in both phases due to the high velocity gradients at the interface between the two phases in free surface flows when differential eddy viscosity models are used. Among the available turbulence models described in the literature, the standard $k-\varepsilon$ turbulence model available in ANSYS Fluent® is chosen for this study. This model is applicable to fully turbulent flow which makes it suitable for the current study for the range of flow rates to be used. Based on the above model verification, the Eulerian-Eulerian Multi-Fluid VOF scheme combined with the standard $k-\varepsilon$ turbulence model can accurately simulate the gas–oil-water three-phase flow.

Mesh and boundary condition

The computational mesh shown as Fig. 1c, which is designed to be dense near the tube wall. In that way, the liquid flow behavior close to the wall and the interface of gas and liquid can be captured accurately. In order to meet the requirements of standard $k-\varepsilon$ turbulence model, the zone close to the wall should be special treated to describe the turbulence flow reasonably. It is applicable to set the $y+ <50$ when take the viscous effects into consideration [17]. Three grids were used for the grids' independency simulation. The pressure gradient is served as the evaluation key index. Table 2 shows the results of grids' independency tests. It is reasonable to select the M 2 as the object, which both can meets the accuracy and saves calculation times.

Table 2

The computing grid parameters

Grid	Grids in computing field	Pressure gradient (kPa/m)
M 1	43164	5.6
M 2	78050	4.8
M 3	127248	4.7

In this paper, the oil, water and gas inlets were all set to velocity inlet boundary conditions, and the outlet of tube was set to outflow boundary condition. An axisymmetric boundary was served as the axis of tube, and the tube wall was set to no-slip boundary condition. To ensure the annular flow over all the cases in this study, the gas and liquid superficial velocities are set as 0.01~0.2 m/s, and 12~ 40 m/s, respectively. The corresponding flow pattern is shown in Fig. 8 [18].

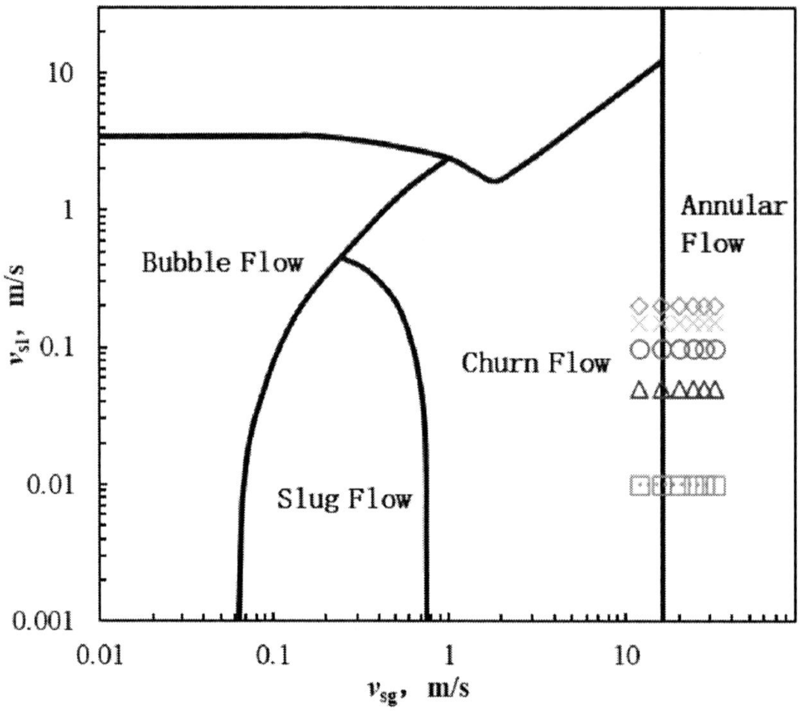

Fig. 3. The test point on the flow pattern map

Model validation

Experimental setup

The experiments were conducted in the multiphase flow loop of the State Key Laboratory of Oil and Gas Reservoir Geology at Southwest Petroleum University (Fig 4). Herein, previous experimental equipment [19,20] were modified. The output pressure of the new compressor could reach 4 MPa, and the maximum gas flow rate was 10 m^3/min. Two 2 m^3 gas storage tanks make the gas flow rate more stable and decrease errors due to fluctuation. The new test loop can be automatically adjusted at any degree from 0° to 90°.

To observe the flow patterns, the tube materials are made up with plexiglass. The experimental visual equipment consists of five Perspex tubes (2-m length and 62-mm inner diameter). The flow rate was measured by an air meter with an accuracy of 1%. A needle valve was installed upstream the air meter to control the air flow rate at any desired value. Tap water and condensate oil was stored in a water and oil tank, respectively. It was supplied by two pumps for the experiments. The liquid flow rate was measured using a turbine meter with a range of 0–0.3 m^3/h and an accuracy of 2%. It was regulated by a valve connecting the turbine meter. The pressures were measured by four pressure sensors in a range of 0–4 MPa and an accuracy of 0.5%.

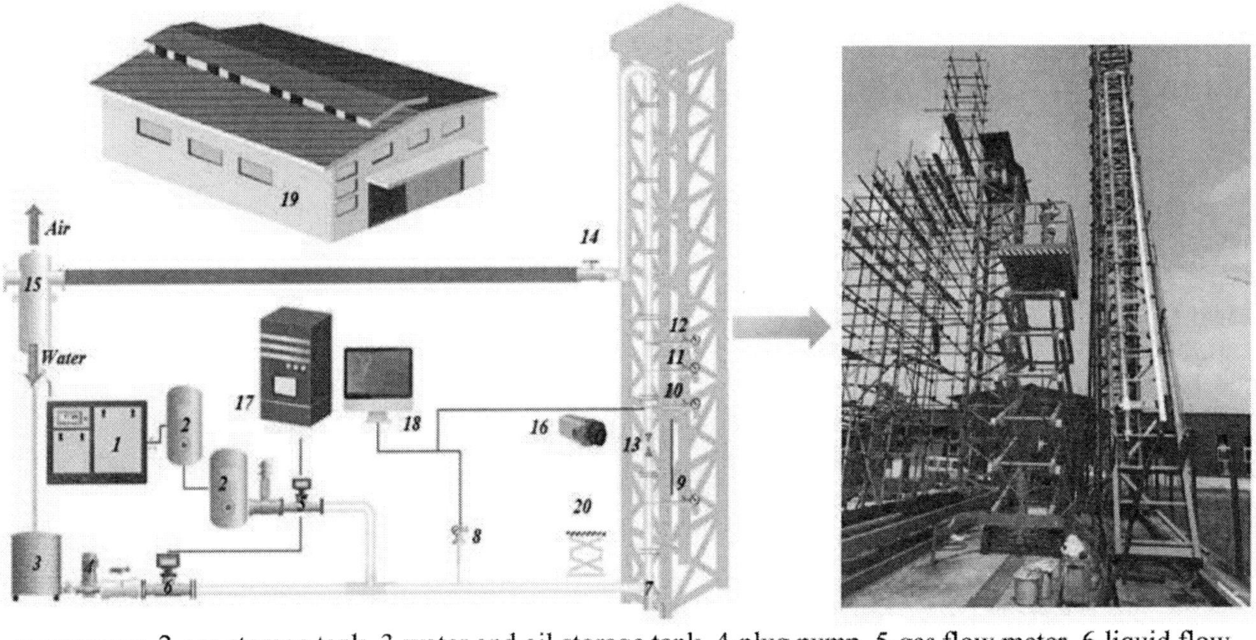

1-compressor, 2-gas storage tank, 3-water and oil storage tank, 4-plug pump, 5-gas flow meter, 6-liquid flow meter, 7-gas-liquid mixing section, 8-bottom pressure gauge, 9 -Pressure gauge upstream choke, 10、11、12-pressure gauge downstream the choke, 13-quick open valve, 14- back pressure valve, 15-gas-liquid separator, 16 -High-speed camera, 17-data recorder, 18-computer, 19-warehouse, 20-lift

Fig. 4. Experimental set up.

Experimental phenomenon discussion

In order to investigate the effect of gas velocity on the phenomenon of oil-water emulsification experimentally, four sets of oil-water-gas three phase flow experiments were conducted in this work. When the U_{SL}=0.1m/s, U_{SL}=0.05,0.5,1 and 16m/s, fw=50%, the experimental snapshots shown in Fig. 5. It is can be observed from the Fig. 5a, the bubble flow pattern can hardly affect on oil-water emulsification due to the low gas velocity. Under the slug flow condition, with the increase of the apparent flow velocity of the gas phase. Owing to the stirring effect of turbulent flow, resulting the phenomenon of oil-water mixture be enforced, as shown in Fig. 5b. With the superficial gas velocity increase to 1m/s, the flow patter of oil-water-gas three phase turned into churn flow. It is obviously to see in Fig. 5c, the interaction of stirring between the oil and water makes the physical properties of liquid changed, resulting the color of in experimental snapshots lighter than before. Fig. 5d shown the phenomenon of oil-water-gas flow under the annular flow condition. The distribution of liquid film attach on the pipe wall is uneven because of the oil phase change the viscosity of the mixed liquid.

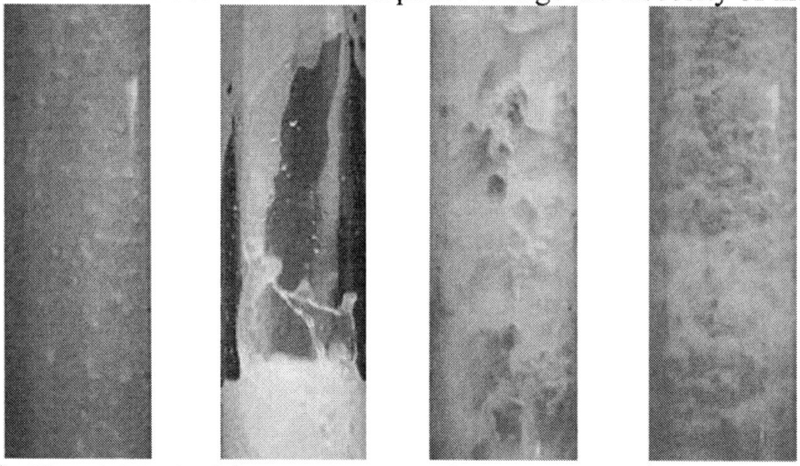

a) U_{SG}=0.05m/s b) U_{SG}=0.5m/s c) U_{SG}=1m/s d) U_{SG}=16m/s

Fig. 5. Experimental observation results

Comparison of Experimental results to Numerical results

In this experiment, it is important to accurately measurement of the corresponding gas flow rate when the liquid film starts to reversal. Based on the liquid film reversal theory, the drag force of the liquid film upward flow is provided by the gas core, and the maximum drag force is at the gas-liquid inner interface. Therefore, as the gas flow rate decreases, the earliest reversal position of the liquid film should be attached to the pipe wall. At present, the most widely used and simplest method to determine the liquid film inversion in experiments is to track the flow direction of the liquid film with a high-speed camera. When the liquid film starts to flow downwards, it can be determined that the liquid film is inverted. At the same time, recording the corresponding gas flow rate is the critical superficial gas velocity.

Some experimental works were conducted to make a validation of simulation results to compare with experimental results under the condition: superficial liquid velocities range from 0.01–0.1 m/s and the fw are 50% and 100%. The critical superficial gas velocities obtained using numerical and experimental data were compared which presented in Fig. 6. It is easy to see that the results agree well under the fw=100% condition, all the errors are below 6%. However, when the fw=50%, the differences between the numerical and experimental results shown the huge gap, and the biggest error increased to 14.89%. This is because oil-water emulsification in experimental study which makes it more difficult for the gas phase to carry the emulsified liquid phase, which makes more liquid phase "stay" in the wellbore, which increases the liquid holdup in the wellbore and further increases the critical superficial gas velocity. The numerical method ignores the effect of oil-water emulsification, and the density of oil is lower than water, which lead to the further error.

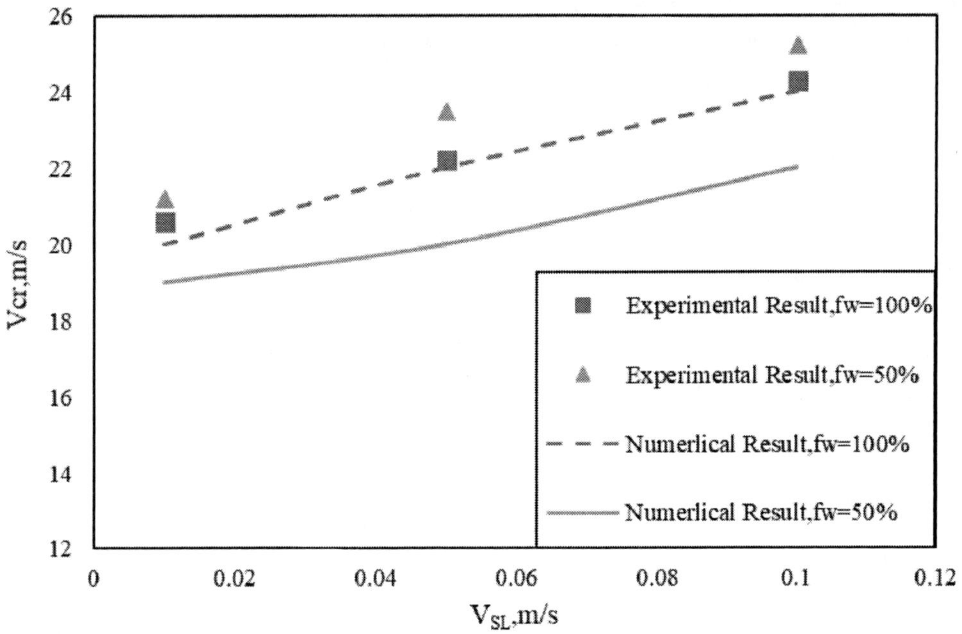

Fig. 6. The comparison analysis of experimental and simulation results

Results and discussion

Typical flow field

Three types of typical flow field are show in Fig. 7, all the case shown a typical annular flow pattern. When the U_{SL}=0.2 m/s and U_{SG}=34 m/s, the liquid phase and streamlines distribution as shown in Fig. 7a. Owing to the high gas velocity, the gas core can provide enough drag force make the liquid film flow upward along the pipe wall. All simulated cases with comparatively high gas velocities show that all the liquid film moves upward out of the pipe. Fig. 7b presents the liquid phase distribution and streamlines when the U_{SG} decrease to 18 m/s, the drag fore provide from gas gore cannot lift all the liquid out of the exit of tube. From the result of streamlines, there are some circle flow fields near the wall, resulting the liquid film in this zoon moves downward. It is means that the velocity is corresponding to the onset of liquid reversal, and also writing as critical superficial gas velocity. As for the condition of U_{SG}=12 m/s, the liquid film attach on the pipe all has been reversal completely. It also means that the drag force from gas core cannot take the film flow upward anymore, and the thickness of liquid film became uneven. From the streamlines show in Fig. 7c, it is can observed clearly that all the liquid film moves downward near the wall periodically.

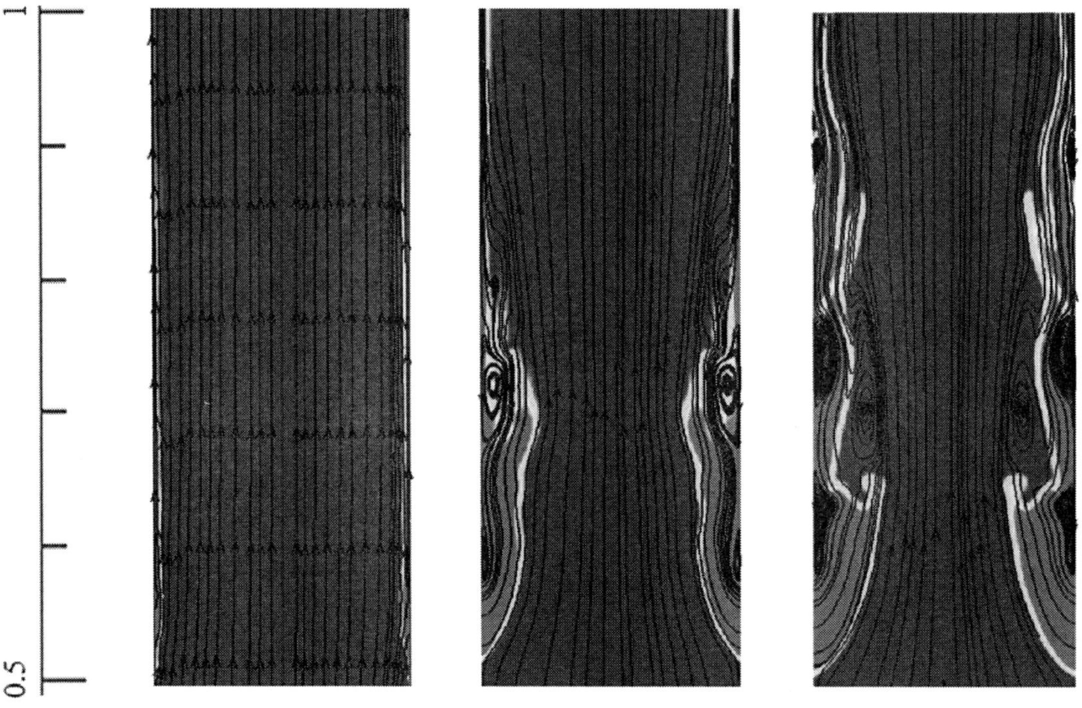

a. U_{SG}= 34m/s b. U_{SG}=18m/s c. U_{SG}=12m/s

Fig. 7. Typical flow field (U_{SL}=0.2 m/s)

Effects of gas superficial velocities

Fig. 8 shows the liquid phase distribution of three different U_{SG} under the same superficial liquid velocity, and the pipe is divided into three parts (0.5m~1m: inlet setion; 1m~1.5m: middle section; 1.5m~2m: outlet section) to better understand the flow characteristics. In this work, it was unable to track the gas bubble interact into the liquid and the liquid droplet entrain in the gas core. As show in

Fig. 8a, the uneven liquid film only take place at the inlet section. That is because the gas inlet is below the liquid inlet, which makes the high-speed gas flow rolls some small waves at the liquid inlet section. Owing to the U_{SL}=28m/s, which above the critical superficial gas velocity, resulting the liquid file of all three sections are all flow upward. When the superficial gas velocity reduced to 18 m/s, the liquid phase distribution of three sections is shown in Fig. 8b. The liquid film of outlet section moves upward near the wall, and shown an even sharp. However, it is clearly to see that the other two sections are all observed the uneven liquid film. Due to the decrease of the gas flow rate, the liquid film starts to reversal at the inlet section. Fig. 8c presents that the whole oil-water-gas three phase flow zoon, there are some periodical roll waves can be observed close to the wall. Even the liquid film of outlet section has been reversal when the superficial gas velocity equal to 12 m/s, which refers to the liquid film reversal completely along the pipe wall.

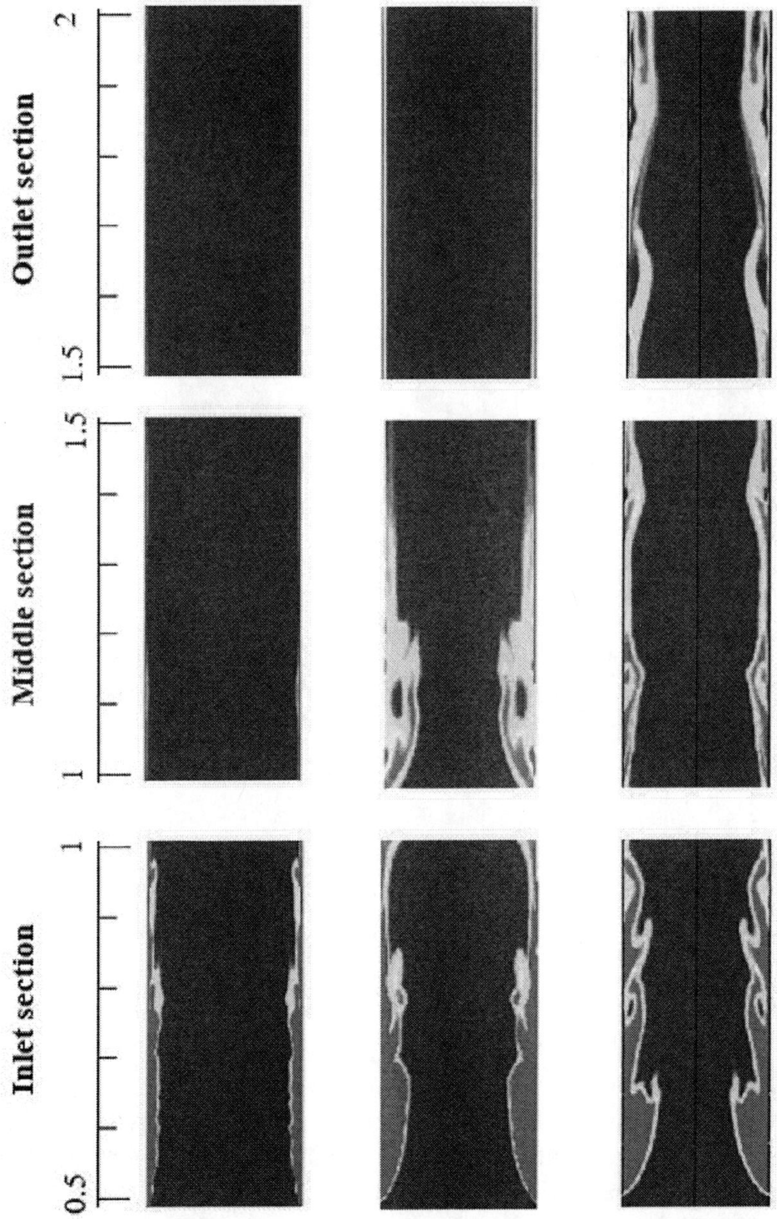

a. U_{SG}= 28m/s b. U_{SG}=18m/s c. U_{SG}=12m/s

Fig. 8. The 2D graphic of liquid phase distribution

Effects of liquid superficial velocities

Under the f_w=50% condition, five superficial liquid velocities were studied to investigate the effect of liquid flow rate on the critical superficial gas velocity. As shown in Fig. 9, critical superficial gas velocity increased with the increase of superficial liquid velocity. From the contours of liquid phase insert in the Fig. 9, it is easy to observe that the thickness of liquid film close to the pipe wall also increased with the increasement of superficial liquid phase. Based on the theory of liquid film reversal, only when the internal shear force below the gravity of liquid film, the phenomenon of reversal take place on the film near the wall. Owing to the high liquid velocity increasing the thickness of liquid film, resulting the higher gravity. Therefore, the higher superficial liquid velocity is required to provide the enough drag force to keep the liquid film moving upward.

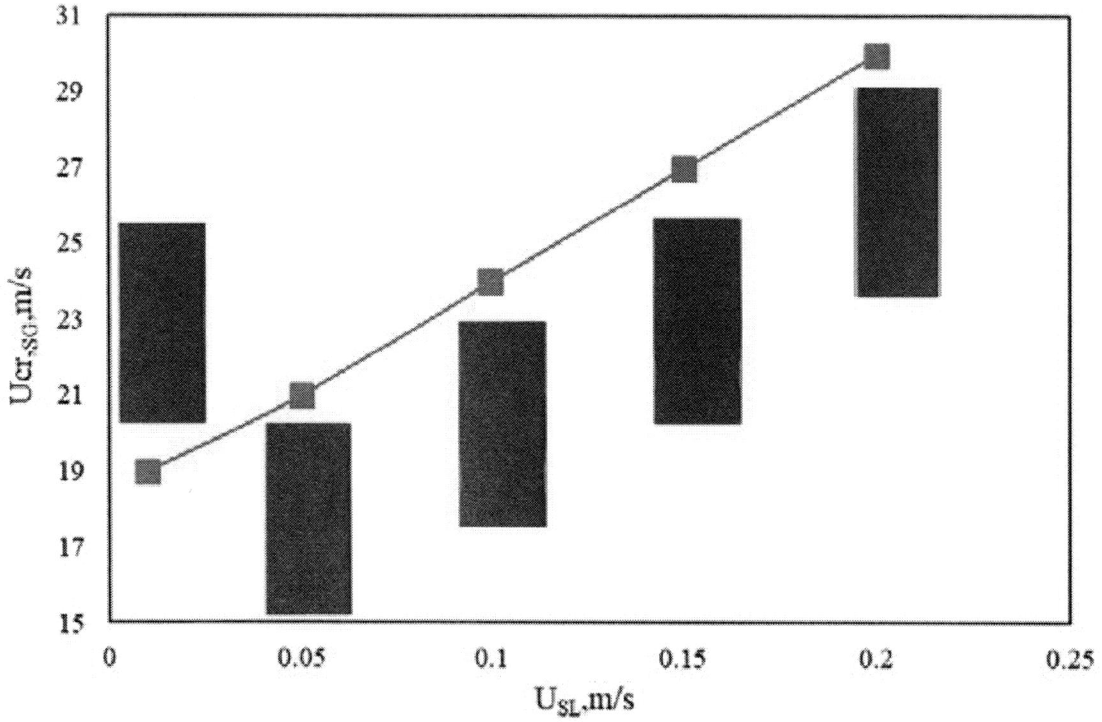

Fig. 9. The critical superficial gas velocity varies with superficial liquid velocity

Effects of pipe diameter

In order to study the effect of pipe diameter on the critical gas flow velocity, when the U_{SL}=0.1m/s and f_w=50%, five different widely discussed pipe diameters simulated in this paper. Fig. 10 presents the relationship between the critical superficial gas velocity and pipe diameter. With the increase of pipe diameter, the critical superficial gas velocity also increased. This is because the larger pipe diameter needs higher gas velocity to drag the liquid film moves upward along the pipe wall. It is also proved by WALLIS, they proposed the correlation about the as-liquid internal shear force, which indicated that the pipe diameter is inversely proportional to the internal friction. Under the same thickness of liquid film condition, the larger the pipe diameter with smaller internal friction and the lower the drag force provided by gas core, resulting the higher critical superficial gas velocity.

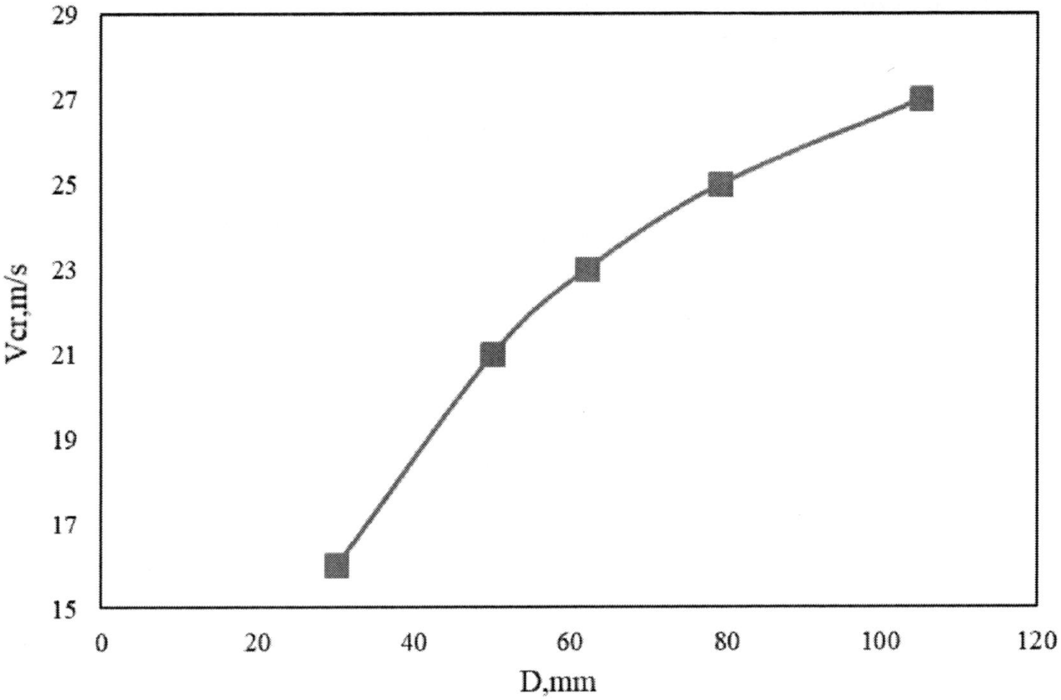

Fig. 10. The critical superficial gas velocity varies with pipe diameter

Effects of f_w

How the condensate oil effect on the critical superficial gas velocity is the most important work in this paper. Therefore, the five sets of fw (0, 25%, 50%, 75% and 100%) were simulated with same pipe diameter and superficial liquid velocity. When the fw=0, all the liquid phase are consisting of condensate oil. Owing to the density of oil is lower than water, resulting the lower drag force needs to maintain the oil film flow upward. On the contrary, the fw=100% means that all the liquid film is occupied by water. In that war, the density of liquid increased in this condition. Basically, it needs more drag force to keep the liquid film moves upward. However, under the annular flow condition, the liquid holdup is much lower, so the effect of viscosity is much higher than gravity of liquid. Comparing to the oil-gas two phase flow (fw=0), the critical superficial gas velocity of water-gas flow is higher slightly, as shown in Fig. 11.

With the fw increased to 25%, the inside of the pipe experiencing oil-water-gas three phase annular flow. Although most part of the liquid film is occupied by oil phase, the small part of water not only can increase the density of liquid, also resulting the oil-water emulsification on the liquid film. The phenomenon of emulsification increases the viscosity of liquid, resulting the critical superficial gas velocity is higher than two phase flow conditions (fw=0 and 100%). Fig. 11 indicates that the highest critical gas velocity is required when the fw=75% in this study. In this condition, it can be seen from the contours illustrate in Fig. 11, the oil-water emulsification forms a matrix slip layer adheres on the pipe wall, which is difficult to flow and blocking the external liquid film. The connection of the external liquid film and matrix slip layer, resulting the difficulty for the gas phase to carry the liquid film upward, which makes the liquid film more likely to reverse downward.

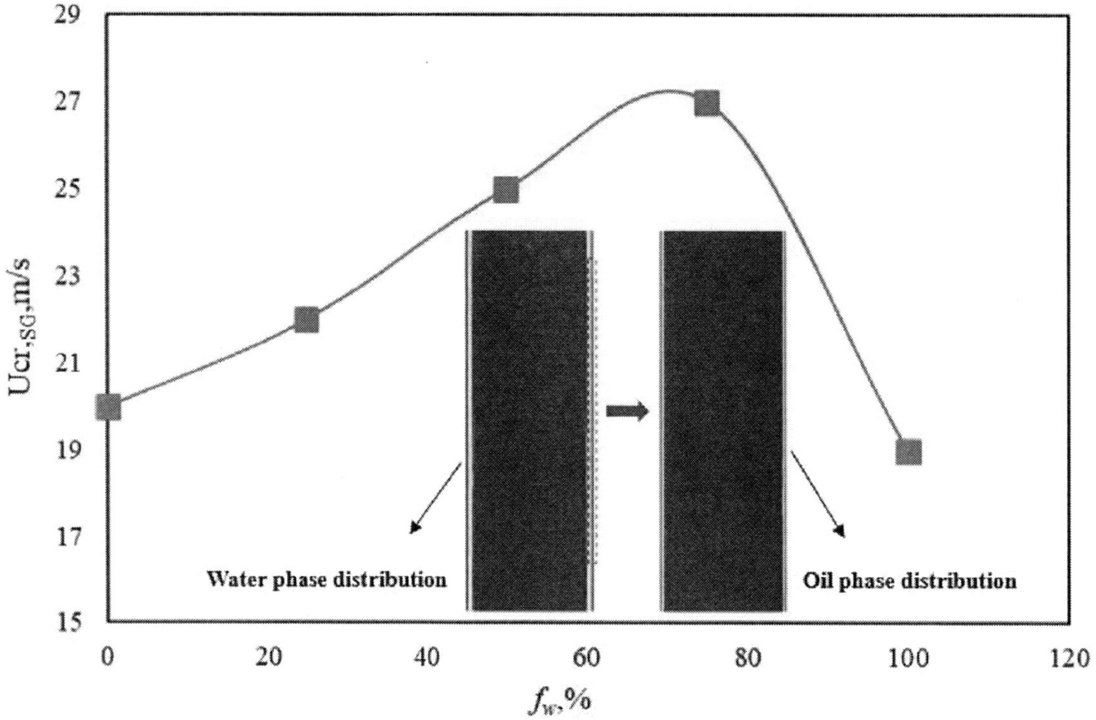

Fig. 11. The critical superficial gas velocity varies with f_w

Effects of slip velocity

Fig. 12 shows that the slip velocity between the gas phase and liquid phase at the x=1500 mm, when the U_{SL}=0.1 m/s. Owing to the high velocity of gas core, the slip phenomenon can be neglected in this region. However, it is can observed that the slip take place at the red circle zoon (liquid film section) in Fig. 12, which the gas velocity and liquid velocity shows an obviously difference in this region. The oil and water velocity profiles at x=1500 mm shows in Fig. 13. There is no slip phenomenon can be observed in the both gas core section and liquid film section. From the Fig. 12 and 13, the velocity profiles show that the slip between the water phase and oil phase can be neglected near the walls, but there is a significant slip phenomenon between gas phase and liquid phase in this region.

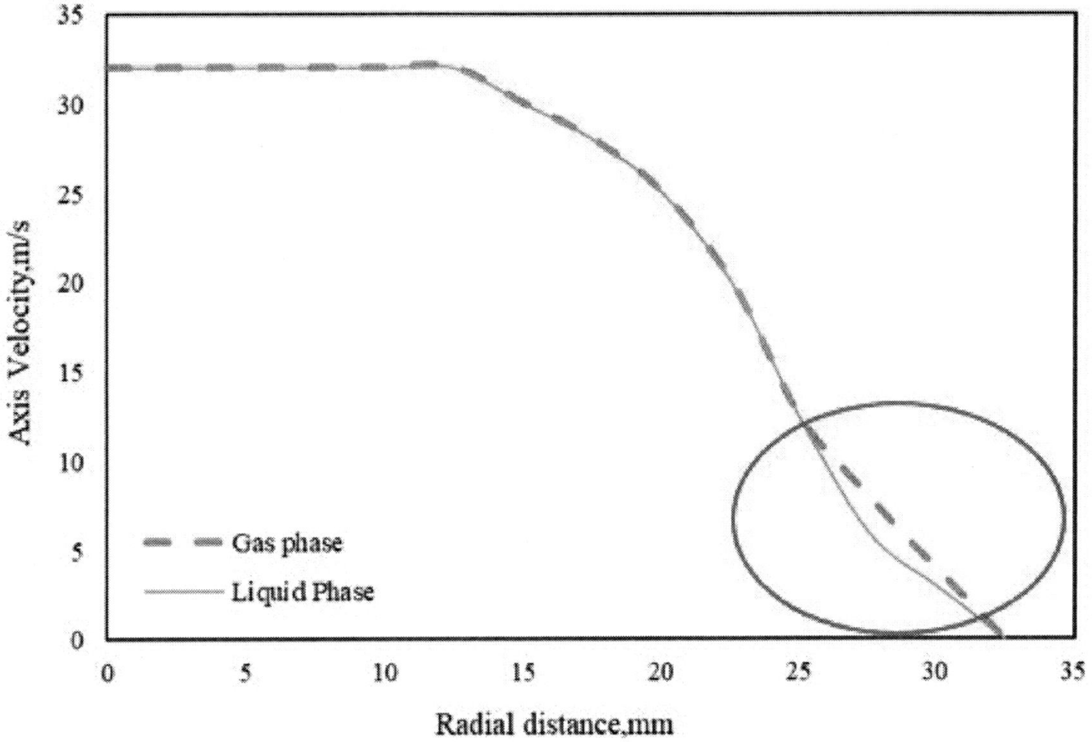

Fig. 12. X=1500mm gas and liquid phase velocity for U_{SL}=1 m/s.

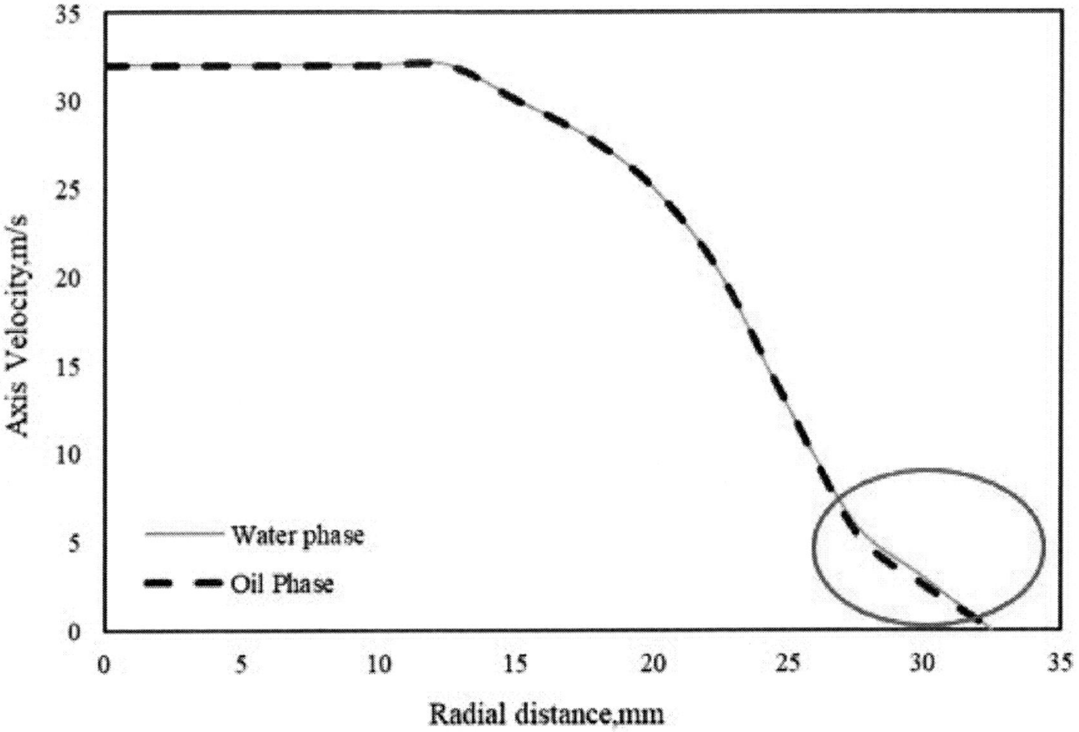

Fig. 13. X=1500mm water and oil phase velocity for U_{SL}=1 m/s.

Conclusions

According to the experimental and numerical studies conducted in this work, several observations and conclusions can be extracted from this study as below.

(1) Compared with the experimental data of critical velocity in the literature, the absolutely error of simulations was below 15% without considering emulsification of oil-water.

(2) Under the oil-water-gas three phase flow conditions, due to the phenomenon of emulsification increases the viscosity of liquid, resulting the critical superficial gas velocity is higher than two phase flow conditions (f_w=0 and 100%).

(3) Based on the numerical results, the highest critical gas velocity is required when the f_w=75% in this study.

(4) The velocity profiles show that the slip between the water phase and oil phase can be neglected near the walls, but there is a significant slip phenomenon between gas phase and liquid phase in this region.

Acknowledgements

The authors gratefully acknowledge the support given by the Science and Technology Cooperation Project of the CNPC-SWPU Innovation Alliance (No.20200CX020203) and China Scholarship Council (No.202009225009).

References

1. Liu Yonghui, Luo Chengcheng, Zhang Liehui, et al. Experimental and Modeling Studies on the Prediction of Liquid Loading onset in Gas Wells[J]. Journal of Natural Gas Science and Engineering, 2018, 57(9): 349-358 .
2. Turner R G, Hubbard M G, Dukler A E. Analysis and Prediction of Minimum Flow Rate for the Continuous Removal of Liquid from Gas Wells [J]. Journal of Petroleum Technology, 1969, 21(11): 1475-1482.
3. Coleman S B, Clay H B, Mccurdy D G, et al. A New Look at Predicting Gas-Well Load-Up [J]. Journal of Petroleum Technology, 1991, 43(3): 329-333.
4. Nosseir M A, Darwich T A, Sayyouh M H, et al. A New Approach for Accurate Prediction of Loading in Gas Wells under Different Flowing Conditions [J]. SPE Production & Facilities, 1997, 15(4): 241-246.
5. Li, J., Almudairis, F., & Zhang, H. Q. (2014, December). Prediction of critical gas velocity of liquid unloading for entire well deviation. In International petroleum technology conference. OnePetro.
6. Guo B, Ghalambor A, Xu C. A Systematic Approach to Predicting Liquid Loading in Gas Wells [J]. SPE Production & Operations, 2006, 21(1): 81-88.
7. Wang, Z., Guo, L., Zhu, S., & Nydal, O. J. (2018). Prediction of the critical gas velocity of liquid unloading in a horizontal gas well. SPE Journal, 23(02), 328-345.
8. Van't Westende, J. M. C., Kemp, H. K., Belt, R. J., Portela, L. M., Mudde, R. F., & Oliemans, R. V. A. (2007). On the role of droplets in cocurrent annular and churn-annular pipe flow. International journal of multiphase flow, 33(6), 595-615.
9. Sawant, P., Ishii, M., & Mori, M. (2008). Droplet entrainment correlation in vertical upward co-current annular two-phase flow. Nuclear Engineering and Design, 238(6), 1342-1352.
10. Alamu, M. B. (2012). Gas-well liquid loading probed with advanced instrumentation. SPE Journal, 17(01), 251-270.
11. Vieiro, J., Asuaje, M., & Polanco, G. (2013). Study of the two-phase liquid loading phenomenon by applying CFD techniques. The International Journal of Multiphysics, 7(4), 301-312.
12. Van't Westende, J. M. C. (2008). Droplets in annular-dispersed gas-liquid pipe-flows.
13. Shiller, L. and Naumann, A. (1935) A Drag Coefficient Correlation. Zeitschrift des Vereins Deutscher Ingenieure, 77, 318-320.
14. Brackbill, J. U., Kothe, D. B., & Zemach, C. (1992). A continuum method for modeling surface tension. Journal of computational physics, 100(2), 335-354.
15. Chen, J., Tang, Y., Zhang, W., Wang, Y., Qiu, L., & Zhang, X. (2015). Computational fluid dynamic simulations on liquid film behaviors at flooding in an inclined pipe. Chinese Journal of Chemical Engineering, 23(9), 1460-1468.
16. Parsi, M., Vieira, R. E., Torres, C. F., Kesana, N. R., McLaury, B. S., Shirazi, S. A., ... & Hampel, U. (2015). On the effect of liquid viscosity on interfacial structures within churn flow: experimental study using wire mesh sensor. Chemical Engineering Science, 130, 221-238.
17. Pope, S. B. (2001). Turbulent flows.
18. Taitel, Y., Bornea, D., & Dukler, A. E. (1980). Modelling flow pattern transitions for steady upward gas-liquid flow in vertical tubes. AIChE Journal, 26(3), 345-354.
19. Liu Yonghui, Luo Chengcheng, Zhang Liehui, et al. Experimental Investigation of the Surfactant Effect on Liquid Removal in Vertical Pipes [J]. Journal of Petroleum Science and Engineering, 2020, 185.

20. Liu, Y., Xie, C., Li, X., Lu, G., Tang, L., & Tan, X. (2020). Optimization analysis of structure parameters of downhole vortex tool based on CFD and orthogonal studies. Journal of Petroleum Science and Engineering, 190, 107030.

630613 Bog Management during Cargo Unloading at the LNG Regasification Terminal

Abstract:

Unloading of LNG cargo from carrier ship to the terminal tank involves passing through loading/discharge arms, jetty lines, rundown lines, and storage tank vertical risers. These lengths are typically made up from several hundred meters to few kilometers and cause inevitable heat leak to the LNG cargo that results in BOG generation. A trade-off exists between unloading flowrate and BOG generated, a higher unloading flowrate delivers the cargo in shorter period results in less demurrage cost. However, a higher unloading flowrate needed a high pumping cost moreover the unloading line temperature also becomes smaller results in a higher ambient heat leak that produces higher BOG. Thus, with the help of a dynamics simulation model of the LNG receiving terminal, a parcel size of 260,000 m^3 is considered for unloading. The study investigated all the ambient factors for devising an optimal unloading strategy. It was mainly found that the storage tank pressure provides the foremost opportunity for optimizing the unloading process that helps best manage the produced BOG.

628216 Integrated Cryogenic Upgradation Process for Baseload LNG Plant

Abstract:

Natural gas (NG) is the cleanest fossil fuel and is regarded as the bridge for the transition from non-renewable to renewable energy sources. Transport of natural gas over long distances as liquified natural gas (LNG) at -160°C and atmospheric pressure, is the most common and economical option. The raw natural gas feed from the field undergoes a series of energy-intensive up-gradation and liquefaction processes to produce the final LNG product, falling in with customer specifications. A realistic baseload LNG plant consists of hot and cold sections, with the former consisting of inlet processing, acid gas removal (AGRU), dehydration, and sulfur removal units, and the latter, operating at sub-ambient temperature, consists of natural gas liquid (NGL) recovery, liquefaction, nitrogen removal (NR), fractionation units, and storage tank.[1] The various processes produce hydrocarbon-rich waste streams, utilized as fuel gas (FG) in a gas turbine to compensate for the power requirements of the LNG plant. This study focuses on incorporating the acid gas removal unit in the cold section by solidifying the acid gases at freezing conditions. It aims to design and optimize an energy-integrated and self-sustained cryogenic process incorporating the AGRU, liquefaction unit, NGL recovery unit, NRU, and export terminal.

The present literature has studied the potential capital and energy saving by removing solidifiable acid gases from process gas stream by direct contact with a cold liquid. [2] The controlled freezing zone (CFZ^{TM}) process, patented by ExxonMobil, consists of an open section in a distillation column, operating at desublimation conditions for solidifying acid gases.[3] The NG vapor containing acid gases moves up through the open section of the column, where it is contacted with liquid methane that has been sprayed through nozzles, causing the acid gases to solidify and methane to vaporize. P S Northrop et al [4] discussed integrated NRU/CFZ^{TM} process operation at 40 bar for feed with a high acid gas content of 65mol% and obtained both pipeline and LNG quality product with acid gas concentration of <2mol% and <50ppm respectively. However, the CFZ^{TM} process for low acid gas content in feed (5 mol%) demands lower operating pressure (10-20 bar) to meet LNG specifications. The energy-intensive liquefaction of this low-pressure NG stream post-AGRU further increases the total energy consumption of the plant and hence the cryogenic AGRU for low acid gas feed to meet LNG specifications remain unexplored.

The inclusion of AGRU in the cold section results in increased cold energy requirements and impacts the operating parameters of downstream processes including NGL recovery, NR, and end flash. The significance of managing boil-off gases (BOG), and balancing the FG streams generated from various processes against the total fuel requirement of the entire plant remain briefly addressed. Although many studies are addressing the two aforementioned problems exclusively, to the best of our knowledge, there is no literature discussing an integrated cryogenic process consisting of AGRU, liquefaction, NGL recovery, and NRU with BOG and hydrocarbon waste stream generation at various process steps accounting for the FG.

In this study, first, we investigate the feasibility of solidifying acid gases from a feed with low acid gas content (5mol%) to meet LNG specifications by incorporating AGRU in the cold section and operating in the freezing line. The AGRU process is integrated with other cold section processes like NGL recovery, NR, end flash, and the export terminal. The cold energy is provided by a refrigerant cycle and a parity between total FG requirement and generation in various processes is established. The NRU consists of one high-pressure column and one low-pressure column, developed in our previous study. The process simulation and thermodynamic calculations are carried out in Hysys, Petro-SIM, and MultiFlash. The feed components and fluid package are selected in MultiFlash. The fluid package selected is GERG-2008 for high accuracy and hydrocarbon components and acid gases are declared as freeze-out components. The material streams and equipment in Petro-SIM support the solid phase and hence the conditions for solidification of acid gases were detected in Petro-SIM. The thermodynamic calculations pertaining to solid formations were carried out by importing the declared MultiFlash file in Petro-SIM. The entire process was simulated in Hysys and the fluid package selected was GERG-2008. The simulation software Hysys and PetroSIM were connected to MATLAB using ActiveX control to allow the exchange of thermodynamic calculation.

Then a simulation-based optimization (SBO) approach deploying particle swarm optimization (PSO), coded in MATLAB, was implemented to minimize the specific power requirement of the plant and find optimal operating conditions of the processes while taking into account the various process and product constraints like higher heating value (HHV) of LNG product, NGL Reid vapor pressure (RVP), Wobbe index (WI), and the pressure of FG and parity between FG heating value and fuel requirements. The energy integration of the process requires identification of solid forming streams and separating them from the liquid and vapor flow stream to prevent the solid formation in the multi-stream heat exchanger (MSHE). The cold energy is provided by a refrigerant cycle and hence the stream providing cold energy for freezing out acid gases from liquids needs to be further identified. The stream for exchanging heat was divided into bundles and then converted to a two-stream heat exchanger network for liquid-vapor flow and liquid-solid flow. [5,6] The performance of the integrated cryogenic process, the design was compared with a conventional LNG up-gradation process from literature.

Keywords: AGRU, Cryogenic, Boil-Off Gas, LNG, NRU

References:

1. Katebah, Mary A., Mohamed M. Hussein, AbdurRahman Shazed, Zineb Bouabidi, and Easa I. Al-musleh. "Rigorous Simulation, Energy and Environmental Analysis of an Actual Baseload LNG Supply Chain." *Computers & Chemical Engineering* 141 (October 4, 2020): 106993. https://doi.org/10.1016/j.compchemeng.2020.106993.
2. Kaminsky, Robert D., and Moses Minta. Systems and methods for using cold liquid to remove solidifiable gas components from process gas streams. SG182399A1, filed January 5, 2011, and issued August 30, 2012. https://patents.google.com/patent/SG182399A1/en.
3. Singh, Vikram, Edward J. Grave, Paul Scott Northrop, and Narasimhan Sundaram. Integrated controlled freeze zone (CFZ) tower and dividing wall (DWC) for enhanced hydrocarbon recovery. United States US8312738B2, filed November 20, 2007, and issued November 20, 2012. https://patents.google.com/patent/US8312738B2/en.
4. Northrop, P. Scott, and Jaime A. Valencia. "The CFZ^{TM} Process: A Cryogenic Method for Handling High- CO2 and H2S Gas Reserves and Facilitating Geosequestration of CO2 and Acid Gases." *Energy Procedia* 1, no. 1 (February 2009): 171–77. https://doi.org/10.1016/j.egypro.2009.01.025.
5. Rao, Harsha Nagesh, and Iftekhar A. Karimi. "A Superstructure-Based Model for Multistream Heat Exchanger Design within Flow Sheet Optimization." *AIChE Journal* 63, no. 9 (2017): 3764–77. https://doi.org/10.1002/aic.15714.
6. "Unified Heat Exchanger Network Synthesis via a Stageless Superstructure | Industrial & Engineering Chemistry Research." Accessed April 9, 2021. https://pubs.acs.org/doi/10.1021/acs.iecr.8b04490.

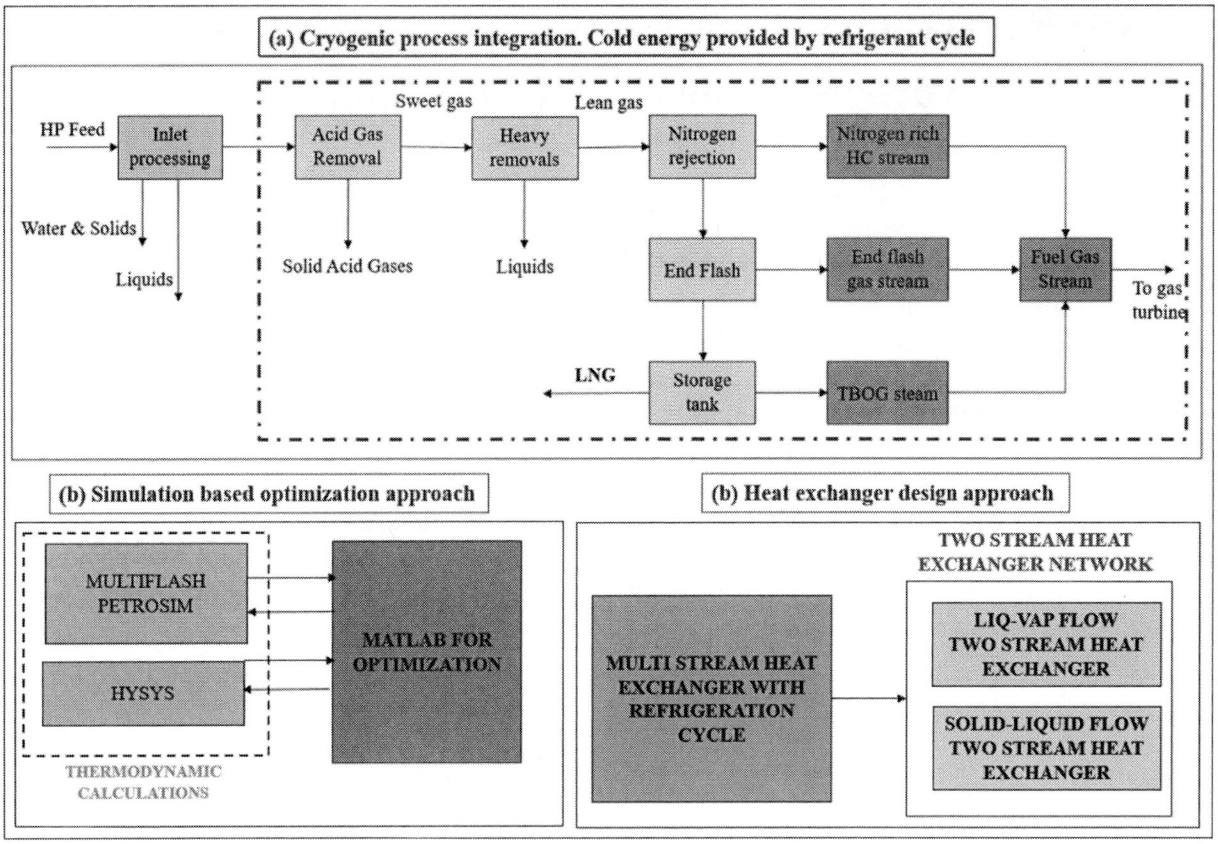
(a) Cryogenic process integration. Cold energy provided by refrigerant cycle
HP Feed
Inlet processing
Water & Solids
Liquids
Sweet gas
Acid Gas Removal
Solid Acid Gases
Lean gas
Heavy removals
Liquids
Nitrogen rejection
Nitrogen rich HC stream
End Flash
End flash gas stream
Fuel Gas Stream
To gas turbine
LNG
Storage tank
TBOG steam
(b) Simulation based optimization approach
MULTIFLASH PETROSIM
HYSYS
MATLAB FOR OPTIMIZATION
THERMODYNAMIC CALCULATIONS
(b) Heat exchanger design approach
MULTI STREAM HEAT EXCHANGER WITH REFRIGERATION CYCLE
TWO STREAM HEAT EXCHANGER NETWORK
LIQ-VAP FLOW TWO STREAM HEAT EXCHANGER
SOLID-LIQUID FLOW TWO STREAM HEAT EXCHANGER

632989 Thermally Induced Two-Step Chemical Looping Process for Hydrogen, Syngas, and Ammonia Production

Abstract:

Complex mixed oxides (e.g. $M_1M_2M_3Ox$) with different combinations of crystalline phases were investigated for a two-step thermochemical looping process for hydrogen, syngas, and ammonia production. Thermodynamically favorable complex metal oxides were synthesized via surfactant templating assisted sol-gel techniques. Different material morphologies such as core-shell nanoparticles and immobilized nanostructures were synthesized and characterized by powdered x-ray diffraction, SEM/EDX, TEM and BET surface area analyzer. In addition, electrochemical impedance spectroscopy was also performed to understand the oxygen diffusion in a complex crystal structure. For the first time, hydrothermally derived carbon from renewable feedstock was employed for carbothermal reduction step. As calcined and fully characterized material was loaded in a continuous tubular reactor and their thermochemical looping potential at temperatures of 900-1200 ^oC was monitored using online sensors and GC. Results obtained on the synthesis and characterization of complex mixed crystalline phase oxides, and their hydrogen, syngas, and ammonia production capability will be presented.

623590 Natural Gas Pyrolysis: Advantaged Production of Low-Emissions Hydrogen and Carbon Materials

Abstract:

Natural Gas Pyrolysis: Advantaged Production of Low-emissions Hydrogen and Carbon Materials

David Dankworth, Jayashree Kalyanaraman, Anjaneya Sarma Kovvali, Sophie Liu, Ying Liu, **Steven Pyl**, Sumathy Raman, Ramon Strauss, Joshua Willis

ExxonMobil Research and Engineering Company (EMRE), Annandale, NJ

Hydrogen produced with low or zero associated GHG emissions is considered an important energy carrier for more sustainable future energy systems; with a broad range of existing and potential applications in transportation, heating, and power generation and storage.

Commonly considered technologies for the production of low-emissions hydrogen include natural gas reforming combined with CO_2 capture and storage (CCS), and water electrolysis powered by renewable electricity. Natural gas pyrolysis has recently emerged as a promising alternative. Pyrolysis (or thermal decomposition) converts the methane in natural gas into hydrogen and solid carbon ($CH_4 \rightarrow C + 2H_2$). Unlike reforming, the pyrolysis chemistry does not inherently generate CO_2. Pyrolysis could therefore enable the use of natural gas for hydrogen production in places where CCS is not feasible for geological or policy reasons. While handling and storing solid carbon (e.g. via landfill or abandoned coal mines) instead of CO_2 can be considered an advantage over reforming; pyrolysis only produces about half as much hydrogen per unit of natural gas. The economics of pyrolysis technologies may become much stronger if the carbon product is a valuable, marketable material (e.g. carbon black, graphene or carbon nanotubes).

In recent work, the potential of natural gas pyrolysis in a fluidized bed reactor as well as in a molten-salt bubble column reactor[1] has been explored. Both of these pyrolysis technologies produce predominantly amorphous solid carbon, but reactor design and operating conditions can be adjusted to tailor the nature and structure of the carbon product. The use of a plasma reactor was investigated for the production of hydrogen and carbon black[2]. The value of the latter, as an additive in rubbers, could significantly help offset the hydrogen production cost.

While several natural gas pyrolysis technologies hold considerable promise, they are currently only practiced at relatively small scales. Significant scale-up challenges remain to turn pyrolysis into a viable option for large-scale production of low-emissions hydrogen and carbon materials. Continued R&D is required to address these challenges, including improvement of methane conversion efficiency, and development of scalable means of providing the required process heat without negating the potential GHG advantages.

References

[1] Brett Parkinson, Clemens F. Patzschke, Dimitrios Nikolis, Sumathy Raman, Klaus Hellgardt, Molten salt bubble columns for low-carbon hydrogen from CH4 pyrolysis: Mass transfer and carbon

formation mechanisms, Chemical Engineering Journal, 417, 2021,127407

[2] Alan Mašláni, Milan Hrabovský, Petr Křenek, Michal Hlína, Sumathy Raman, Vineet Singh Sikarwar, Michal Jeremiáš, Pyrolysis of methane via thermal steam plasma for the production of hydrogen and carbon black, International Journal of Hydrogen Energy, 46 (2), 2021, 1605-1614

622958 Cellulose As a New Templating Agent for Combustion Synthesis of NiO-MgO Solid Solution Catalysts for the Dry Reforming of Methane

Abstract:

Introduction

Due to the similar crystalline structures and lattice parameters NiO and MgO can form substitutional solid solutions, which, upon reduction, can be an effective catalysts for the dry reforming of methane (DRM). It is known that Ni supported on NiO-MgO solid solution catalysts are quite stable towards carbon formation. This stability is attributed to the strong metal support interactions and the basicity of MgO, which enables the adsorption/activation of carbon dioxide with subsequent oxidation of deposited carbon [1]. However, the preparation methods of NiO-MgO solid solutions usually include prolonged calcination steps at elevated temperatures which decreases the surface area of the obtained powders, resulting in lower catalytic activity [2].

In this work we present a new route to prepare NiO-MgO solid solutions, namely by solution combustion (SC) synthesis method. We showed that it is possible to obtain high surface area NiO-MgO active and stable catalysts when using glycine as a combustion fuel for redox reaction with metal nitrates assisted by cellulose as a templating agent. Time on stream (TOS) studies showed that the obtained catalysts exhibit high DRM activity comparable to the Ni-Pt catalysts reported in the literature, and have enhanced stability towards carbon formation. We found that under certain conditions, cellulose acts as a template and additional fuel, thus resulting in structured catalysts with up to 150 m^2/g BET surface area. The mechanism of combustion synthesis reaction and the main factors responsible for the templating effect and high surface area are also reported.

Materials and Methods

Nickel nitrate and magnesium nitrate hexahydrates, with a ratio 10, 20 and 30 wt. % of Ni in final powders were dissolved in deionized water and impregnated on cellulose paper (Whatman, grade 4 filter paper). Upon drying at 80 ^oC, the impregnated cellulose sheets were ignited to initiate the combustion reaction resulting in the formation of NiO-MgO solid solutions. The resulting powders were first calcined and then reduced with 5% H_2 at 600 ^oC. This treatment segregated some Ni to the surface and resulted in catalysts consisting of Ni supported on NiO-MgO. DRM catalytic activity was measured in the temperature range of 500-700 ^oC in a continuous flow fixed bed quartz reactor at atmospheric pressure (total flow - 120 ml/min, CH_4, CO_2 = 20%, balanced He). The combustion synthesis reaction was monitored by High Speed Infrared Camera, and in-situ TG and DSC measurements, were performed to evaluate the phase formations. BET adsorption experiments and SEM/TEM microscopy were used to determine the key structural features of the catalysts. In addition, catalysts were also characterized by TPR, chemisorption, XRD and XPS methods.

Results and Discussion

The application of cellulose paper during combustion synthesis greatly influences the surface area of the catalysts. The regular solution combustion (SC) resulted in a NiO-MgO solid solutions with a BET surface area of up to 23m^2/g, while the paper assisted (PA) combustion synthesis method allowed to obtain catalysts with up to 140 m^2/g surface area. Upon reduction, smaller Ni crystallites segregated to the surface resulting in a higher catalytic activity. The highest methane conversion was obtained for 10%Ni/NiO-MgO (PA) catalysts, reaching 60% at 600 ^oC and 75% at 700 ^oC.

Figure 1 shows the microstructures of the initial filter paper and the combustion synthesis products. It can be seen that cellulose paper (1a) has a fibrous structure which is reflected on the structure of final powders. The templating effect is more pronounced for NiO (1b) showing fiber-like structure after the combustion reaction, whereas for MgO (1c) the structure is mostly porous. For 10%NiO-MgO solid solution (1d) these two effects combine with each other resulting in a fibrous, porous structures responsible for the high surface area of final catalysts.

Significance

The use of cellulose as a templating fuel enables fast and energy efficient synthesis of active and stable catalysts for the DRM. This method can be used also for the synthesis of high surface area materials for other applications, which include the formation of solid solutions and metal-support complex interactions.

References

1. Ruckenstein, E., Hu, Y.H., Appl. Cat. A 154, 185-205 (1997).
2. Zanganeh, R.M., Rezaei, M., Zamaniyan, A. Adv. Powder Technol. 25, 1111-1117 (2014).

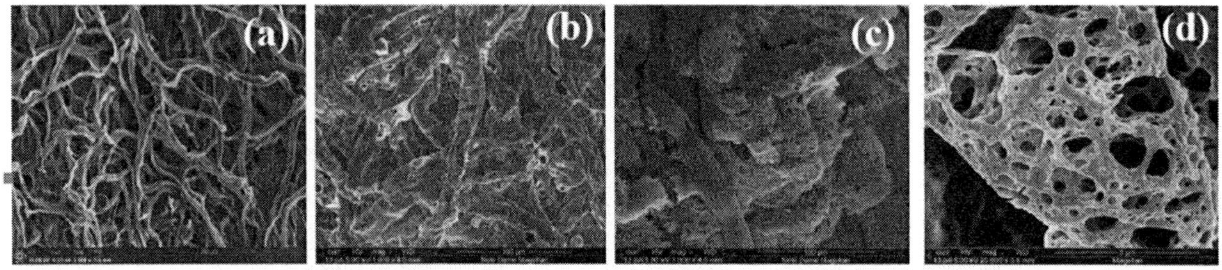

A NOVEL PHOTOCATALYTIC PROCESS FOR ELECTRIFICATION OF HYDROGEN PRODUCTION BY STEAM METHANE REFORMING

Dr. Syed Ali Gardezi, Syzygy Plasmonics, Houston Texas

MOTIVATION

One of the key issues of our time is the dual challenge of meeting the increasing demand for low cost, safe, and reliable energy, while simultaneously lowering the risk of global warming. There is an ever-growing desire to reach net-zero greenhouse gas emissions by mid-century and limit the global temperature rise to 1.5 °C [1]. In this regard, zero-emission energy carriers are becoming increasingly attractive, especially hydrogen **(H_2)** with its wide range of uses. Electrification of hydrogen production via Steam Methane Reforming (SMR) while eliminating greenhouse carbonaceous emissions is made possible via plasmonic photocatalysis. The heart of the process is a novel photocatalytic reactor that uses light-emitting diodes (LED's) to drive the reforming reaction in the presence of a plasmonic nanoparticle photocatalyst.

Plasmonic Catalysis

Metal nanoparticles can couple electromagnetic radiation with electron oscillation. The combined resonance gives rise to an active optical nano-antenna [2]. Recently these nano-antennas have been shown to function as light-driven catalysts that can cause chemical change on their surface. There are two mechanisms (i) phonon-driven or (ii) charge-carrier-driven [2]. The metals that can give rise to either (or both) mechanisms include Au, Ag, Cu, Al [3, 4, 5, 6]. These plasmonic metals although being active to light, are not universally considered as good catalysts. However, more recent work at Rice University has shown that these plasmonic nanoparticles, in addition to enhancing light absorption, can also modify the adjacent metal nanoparticle surface. This is called an antenna-reactor complex, which consists of two very distinct materials [2]. This distinctive coupling generates hot carriers within the catalytic nanoparticle, and thus functions as an efficient light-controlled reactive catalyst. One such "reactor-antenna plasmonic catalyst" is shown in Figure 1.

Figure 1. The electron microscope image of nanoscale antenna-reactor combination [7]

The photocatalytic properties of the antenna–reactor complex are translatable to chemical reactions, such as reforming and hydrogenation. Important and industrially relevant reactions are steam reforming and the selective reduction of acetylene [8]. The former is a commercially viable process for producing hydrogen/syngas (a precursor for marketable products, e.g., fuels, methanol) from greenhouse gases. Likewise, ethylene is a commodity chemical precursor used in the production of polyethylene-based materials with widespread commercial use [8]. One important aspect is that plasmonic-induced hot carriers lead to rapid desorption of hydrogen, biasing equilibrium towards desorption and limiting the availability of hydrogen on the surface, which is favorable to drive the reforming reaction or ethylene hydrogenation forward. Following this trend, it has been shown recently that a Pt–Cu single-atom alloy produce stable intermediate radicals and hydrocarbon products and no coking is observed [9]. Likewise, Cu–Ru single-atom alloy nanoparticles are also coke-resistant plasmonic photocatalysts [10]. The small plasmonic Cu antennas provide efficient light absorption and thus efficient generation of hot carriers under illumination, while single-atom Pt or Ru sites on the Cu surface offer high catalytic activity [9,10]. The local electronic inhomogeneity at the reactor site could further facilitate the generation of hot carrier [10].

Photocatalytic Reactor

Irradiation of photocatalytic process may be achieved by using (i) natural (e.g., solar) or (ii) artificial sources of light (e.g., IR lamps, UV lamps, arc lamps, or LEDs. Typical photo-reactor configurations available today include (i) slurry reactors (ii) annular reactors, (iii) immersion reactors and (iv) optical fiber/tube reactors. [11]. Commonly reported drawbacks associated with these types of reactors include lack of uniform irradiance of the photocatalyst and mass transfer limitations associated with insufficient contact between the photocatalyst and fluid. Further, difficulties in reactor design, material selection, and an incomplete understanding of the critical parameters that are important to reactor design have also hampered prior development efforts.

Figure 2 Photocatalytic reactor employed at Syzygy Plasmonics.

To leverage the benefits of the antenna-reactor photocatalysts, we have developed a novel photocatalytic reactor (Figure 2) to enable the irradiation of the catalyst bed. The gaseous reactants flow through the packed photocatalyst bed as in a plug-flow reactor, allowing for continuous reaction and generation of the desired product. The milder reaction conditions afforded by this technology means that the reactor can be constructed from simple materials, such as quartz glass and aluminum.

System Design

To demonstrate the efficacy of both the photocatalyst and associated reactor, we have designed an integrated system to produce hydrogen from pipeline-grade natural gas (Figure 3) Pipeline-grade natural gas may contain trace amounts of naturally occurring sulfur, as well as sulfur-containing odorants. Sulfur is known to poison to many catalysts. This warrants the use of desulfurization technology to avoid ingress of sulfur into the photocatalytic reactor. The system is designed to remove both organic and inorganic sulfur.

The photocatalytic steam methane reformer (PSMR™) follows the desulfurization process. The enhanced performance of the photocatalyst enables operation at a few hundred degrees Celsius below conventional steam methane reforming, and at less than 100 Psi(g) pressure. The steam methane reforming reaction primarily produces syngas, a mixture carbon monoxide and hydrogen (Reaction 1 below). The PSMR effluent is further reacted in a downstream water gas shift reactor, to convert much of the carbon monoxide to carbon dioxide, along with additional hydrogen as shown in Reaction 2 below.

$$CH_4 + H_2O \longrightarrow CO + 3\,H_2 \qquad (1)$$
$$CO + H_2O \longrightarrow CO_2 + H_2 \qquad (2)$$

Hydrogen is recovered from the product gases using pressure swing adsorption (PSA system). The use of an in-house designed PSA ensures maximum purity of product hydrogen (99%+) and a separation efficiency of approximately 85%. One of the major considerations is to optimize the use of energy in this process. The PSMR demonstration system has been designed to recycle heat from downstream of the reactors to provide heat for the steam generation. As we scale the system up to higher throughputs processes like ORC (organic Rankine cycle) may be employed to recover additional heat and further optimize the electricity usage. A salient feature of this electrification process is the potential elimination of carbon dioxide emissions. A major benefit over the traditional SMR process is the absence of any greenhouse emissions that originate from the combustion of natural gas to heat the reactor; the only energy input required for the PSMR is electricity, which can be supplied by renewable sources.

As far as the bi-product carbon dioxide is concerned, we are currently developing photocatalytic Dry Methane Reforming [12] to react carbon dioxide with methane in a process that can eventually lead to the production of methanol. As such, we have a pathway to make the entire process carbon neutral.

System's Readiness for Deployment

The developed technology is being installed for demonstration; to ensure success several measures are undertaken. This includes (i) following industrial standards for designing the system other than the novel photocatalytic reactor (ii) following the OSHA process safety management (PSM) program to ensure design and operational safety (iii) implementation of automation and controls for optimization and a smooth transition into fail-safe mode in case of any upset (iv) thorough quality control (QC) strategies during procurement, and system design and operation (calibration of instruments, development of KPI's for reporting and plant optimization). Table 1 depicts the full breadth of system design work undertaken during the project development phase that ensures success and reliability in the operational phase. Figure 4 reinforces the robustness of the technology as it depicts a stable performance of our plasmonic photocatalyst during 1000 hours of testing. These strong indicators reinforce the potential of success for the first demonstration unit that is ready for operation.

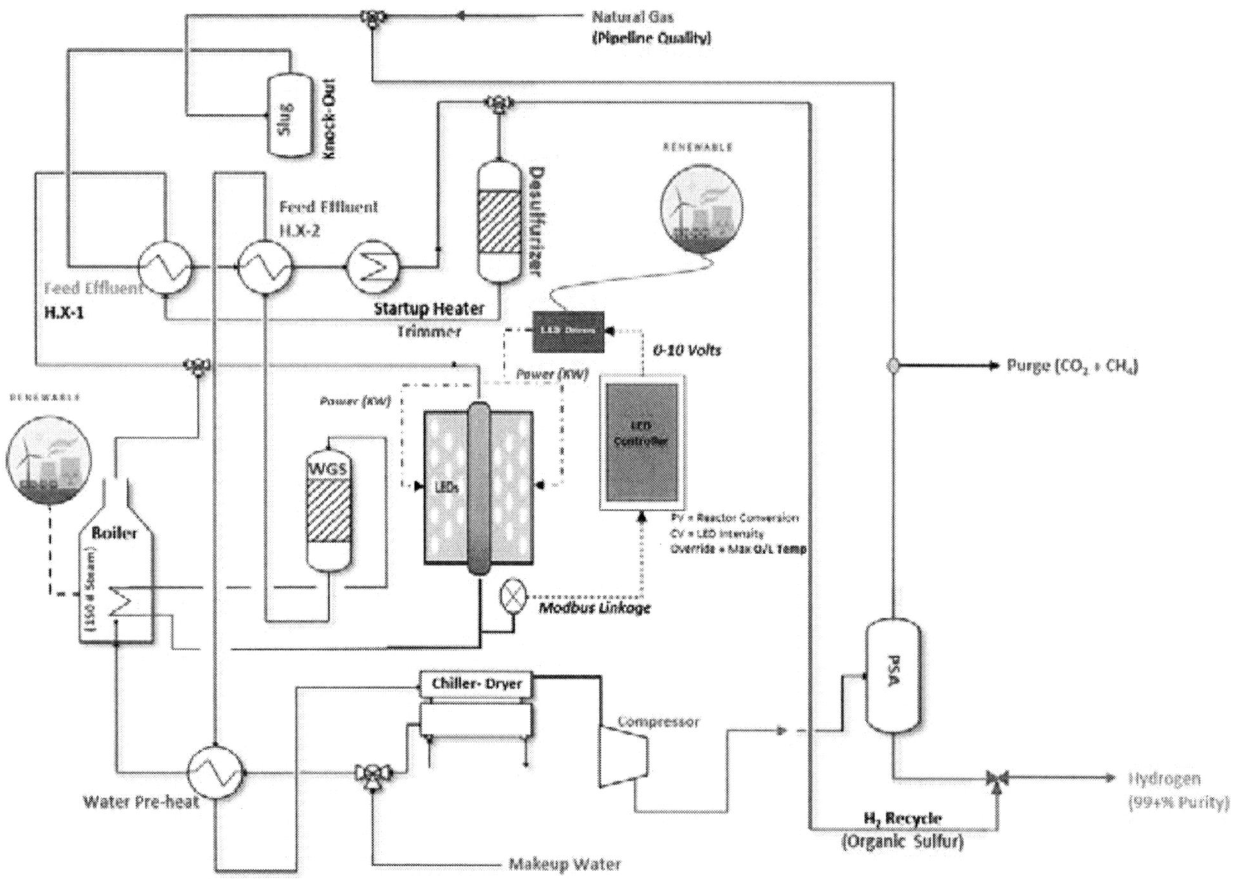

Figure 3: Overview of 5Kg/day hydrogen producing facility using our photocatalytic steam methane reformer

Table 1. Engineering Design Package adopted for successful demonstration of system incorporating plasmonic photocatalytic reactor

Process	Mechanical	Safety	Controls	Electrical	Quality Control	Project Mgmt.	Operations
PFDs	Selected Equipment Specs	Pre-PHA Report	Control Narratives	Power and Instrumentation Diagrams	FAT Requirements	Project Plan	MOC Process
P&IDs	Vendor Drawings	FMEA Report	Interlock Logic Diagrams	Wiring Diagrams	Inspection Test Plan for Major Equipment	Budgetary Estimates	Equipment Operating Manuals
BEDD	Structural Drawings	HAZOP Report	Cause and Effect Matrix		Inspection Procedure		Equipment Operating Procedures
Equipment Specs for Procurement	Assembly BOMs	What-if Report	Controls & Interlock Programing		Calibration Procedures		
Utility Requirements	Sparing lists	Alarm Management					
Process Models							

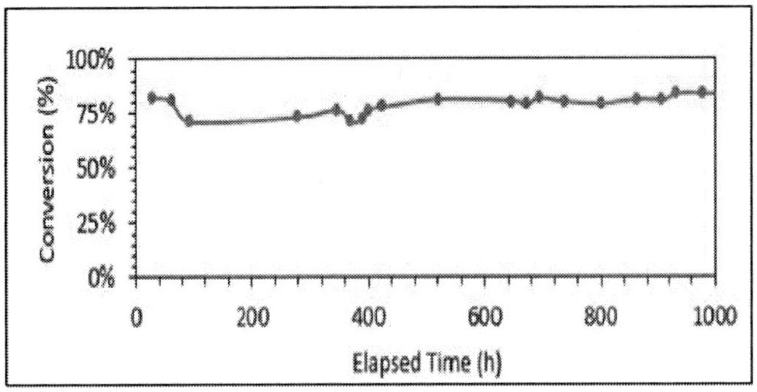

Figure 4: Steady performance of plasmonic catalyst following 1000-hr testing in a photocatalytic reactor at test rig.

CONCLUSION

In conclusion, an integrated system has been developed that incorporates a photocatalytic reactor for producing fuel cell grade hydrogen from natural gas. This photocatalytic reactor has a fixed bed of plasmonic antenna-reactor catalyst. This system employs an in-house designed pressure swing adsorption system for recovering 99%+ purity hydrogen. The rest of the system is designed following established industrial norms. This system will be the first successful demonstration of this technology en-route to commercialization.

REFRENCES

1. "Green Hydrogen, A Guide to Policy Making", International Renewable Energy Agency, 2020.
2. Swearer, D. F. et al. "Heterometallic Antenna-Reactor Complexes for Photocatalysis," Proc. Natl Acad. Sci. USA 113: 8916–8920 (2016)
3. Mukherjee, S. et al. "Hot electrons do the impossible: Plasmon-induced dissociation of H2 on Au," Nano Lett 13(1):240–247 (2013).
4. Xie, W. & Schlücker, S. "Hot electron-induced reduction of small molecules on photo-recycling metal surfaces," Nat Commun 6: 7570 (2015).
5. Marimuthu, A. Zhang, J. & Linic, S. "Tuning selectivity in propylene epoxidation by plasmon mediated photo-switching of Cu oxidation state," Science 339 (6127): 1590–1593 (2013).
6. Zhou, L. et al. "Aluminum nanocrystals as a plasmonic photocatalyst for hydrogen dissociation," Nano Lett 16 (2):1478–1484 (2016).
7. Syzygy Plasmonics website, https://plasmonics.tech/technology-1.
8. Tiedtke, DB. et al. "Chemicals influencing the activity of palladium-based catalysts for the selective hydrogenation of acetylene to ethylene in acetylene converters. Proceedings of the 13th Ethylene Producers' Conference," (Am Inst Chem Eng, New York), pp 586–607 (2001).
9. Marcinkowski, M. D. et al. "Pt/Cu single-atom alloys as coke-resistant catalysts for efficient C–H activation," Nat. Chem. 10, 325–332 (2018).
10. L, Zhou. et al. "Light-driven methane dry reforming with single atomic site antenna-reactor plasmonic photocatalysts," Nat Energy 5: 61-70 (2020)
11. Gerven, V. et al. "A Review of Intensification of Photocatalytic Processes," Chem Eng Process 46*:* 781–789 (2020)
12. "Light-Driven Conversion of CO2 and Methane to Syngas and Methanol," NSF Award # 2052133

627285 Photobacteria-Based Biohydrogen Production from Organic Waste By the Use of Near-IR Light and Plasmonic Nanoparticles

Abstract:

An emerging hydrogen economy is amplifying a demand for clean hydrogen production technologies to better match the diverse applications of hydrogen. While conventional hydrogen applications are supplied by large central-station gas reforming units, emerging applications create a more disparate demand for hydrogen and thus require an innovative solution to the infrastructure and distribution hurdles that large central-station hydrogen production is facing. Our approach involves converting organic wastes to hydrogen gas. The simultaneous elimination of organic wastes and production of clean fuels will have an immense impact to both society and industrial manufacturing sector. Enhanced understanding of the interface between nanoparticles and photo-responsive bacteria (*Rhodopseudomonas palustris)* will further advance knowledge in interactions with biological systems. Although literature shows the production of gases by photobacteria but here we demonstrate the integration of photonics, biology, and nanostructured plasmonic materials for hydrogen production with lower greenhouse CO2 gas content at quantified light energy intensity and wavelength. Phototrophic purple non-sulfur bacteria are able to generate hydrogen as a byproduct of nitrogen fixation using energy absorbed from visible and near-IR (NIR) light. This type of biological hydrogen production has suffered from a low efficiency of converting light energy into hydrogen, in part due to light sources that do not exploit the organisms' capacity for NIR absorption. We used NIR light sources and optically resonant gold-silica core-shell nanoparticles to increase light utilization of the bacteria to convert waste organic acids, such as acetic and maleic acids, to hydrogen. Further, management of the microbes leads to lower CO2 emissions while maximizing hydrogen production. Batch growth studies with Rhodopseudomonas palustris demonstrated > 2.5 fold increases in hydrogen production when grown under a NIR source (167±18 micromol H2) compared to a broad-band light source (60 ±6 μmol H2) at equal light intensity (130 W/m¬2). The addition of optically resonant gold-silica core-shell nanoparticles enhanced near field light utilization and further improved hydrogen production more than two-fold. The presentation will include hydrogen production rates, role of light intensity, use of filtered solar light, and optimization of bioreactor conditions to maximize hydrogen production with lower CO2 content. This work has been funded by the National Science Foundation, NIEHS, and by Southern Company.

632808 The Case of the Missing Asphaltenes from the Naphthenic Froth Treatment Process

Abstract:

Asphaltenes are a complex polydisperse system, and the molecular structure, kinetic and thermodynamic properties vary within a crude oil sample. The effects of this polydispersity can be manifested in various ways. For example, the sizes of asphaltene clusters follow a power-law distribution, where larger asphaltenes exist in decreasing likelihood[1] and the largest are nearest their phase transition[2]. Further, the double bonds equivalent of asphaltene molecules is inversely correlated to asphaltenes solubility[3], and surface-active asphaltenes can be separated by adsorption onto inorganic or water interfaces[4]. Therefore, the isolation and investigation of the most impactful asphaltenes are pivotal for understanding the influence of asphaltenes on different phenomena. In this presentation, we will present evidence that a fraction of asphaltenes is lost during the naphthenic bitumen extraction process used in the oil sands industry. We used toluene to dilute two bitumen samples, one extracted by toluene and the other one extracted by naphtha, then titrated heptane into diluted bitumen solution to destabilize asphaltenes. The naphtha extracted bitumen needs higher heptane concentration to destabilize asphaltenes, which implies the poorly soluble asphaltenes are lost in the naphthenic extraction process. These "missing asphaltenes" are hypothesized to migrate to water-in-oil and inorganic clay interfaces, and are thus removed from the bitumen product. The effect of these processes on stabilizing water-in-oil emulsions is of critical concern to improve product quality and reduce corrosion concerns. Evidence for this hypothesis comes from observations on asphaltenes and solids co-aggregation, where co-aggregation clusters were cut by Focused Ion Beam and imaged by SEM. Nano-scale (20~200 nm) clay particles were found to be distributed in the precipitated asphaltene particles. Based on these findings, our objective is to isolate and investigate the missing asphaltenes, which may be the "tail end" of the size distribution or the poorly soluble asphaltenes. Based on our preliminary results, approximately less than 1% of the asphaltenes (equivalent approx. 0.1% of bitumen mass) are lost in the naphthenic bitumen extraction. To extract and characterize this "missing" fraction, we assembled a nano-filtration system to isolate the most unstable asphaltenes and the nano-scale inorganic solids these asphaltenes may associate with. The filtration process was operated under process conditions with a bitumen (B) to diluent (D, naphtha) ratio of D:B = 0.7 at 80 °C.

626472 Microwave-Assisted Dehydrogenation of Ethane over Stable Csru/CeO_2 Catalyst for Production of Light Olefins

Abstract:

Light olefins such as ethylene, propylene, and butenes which are well known as raw chemicals of the chemical industry, have been widely used to produce fibers, plastics, synthetic rubber, etc. Currently, C_2 and C_3 olefins are mainly made from natural gas via multi-step processes, syngas formation, methanol synthesis and MTO. Moreover, there is no commercial C_4 olefins production process using natural gas as feedstock. Direct nonoxidative dehydrogenation is promising for production of light olefins from natural gas. However, the nonoxidative dehydrogenation suffers from rapid catalyst deactivation resulting from coke formation.

In this study, a novel catalytic process for direct conversion of natural gas to light olefins was developed. Under microwave irradiation, the light olefins were efficiently produced over CsRu/CeO_2 catalyst using ethane as feedstock. The CsRu/CeO_2 catalyst demonstrates extremely high activity and selectivity towards light olefins. For example, at 600 °C and atmosphere pressure, ethane conversion reached to 74%, with 76% light olefins selectivity. Beside C_2 and C_3 olefins, C_4 olefins are also formed. Specially, C_4 olefins consisted of mainly 1,3-butadien which is the raw material for producing synthetic rubber, such as styrene-butadiene rubber, polybutadiene rubber, acrylonitrile–butadiene–styrene, nitrile rubbers, and chloroprene rubber, etc. Moreover, the CsRu/CeO_2 catalyst demonstrates superior stability during the catalytic process. After total 72 hours' time-on-stream test, the CsRu/CeO_2 catalyst still maintained high activity without any sign of deactivation, even with the variation of reaction conditions (temperature and feeding) and several on-purpose shut-down and start-up. This work demonstrates that there is significant potential for cost effective conversion of natural gas to light olefins by microwave catalytic technology.

627358 Thermochemical gas splitting using iron aluminate-based materials

Abstract:

Thermochemical gas splitting of water and/or carbon dioxide via redox is a promising technology to renewably produce syngas for the diversification of gas-to-liquid (GTL) fuels. Here, a metal oxide intermediate undergoes a two-step redox cycle, where either oxygen or hydrogen and/or carbon monoxide is produced. To date, experimental demonstrations have yet to report efficiencies necessary for competing with conventional technologies, primarily due to the insufficient properties of the active materials considered and the requirement for large temperature swings between redox regimes. As a result, recent research has largely focused on discovering new materials with thermodynamic properties that permit lower temperature operation without sacrificing product yields. However, such efforts have proven difficult, as materials that are reducible at low temperatures often become harder to oxidize (and produce fuel).

In terms of efficiency, performing each reaction at the same temperature is appealing due to the elimination of the significant sensible heating penalties that are associated with conventional temperature swing strategies. Although isothermal operation implies lower hydrogen/carbon monoxide yields due to operating the exothermic oxidation reaction under unfavorable conditions, material selection for isothermal shifts to focus on materials that undergo large extents of reaction within the attainable range of oxygen chemical potential.

Recent research has shown that iron aluminate-based materials exhibit uniquely large yields under isothermal conditions, making them ideal candidates for efficient fuel production. Here, a kinetic investigation of these materials is conducted in a stagnation flow reactor and thermogravimetric analyzer to gain insight on how these materials split water and/or carbon dioxide. Results indicate these materials can provide a path forward to a clean, efficient generation of fuels.

629279 Comparison of the Technologies Utilized for Refining of the Utilized Lubricating Oil

Abstract:

The lubricant oil is one of the most utilized oil in any of the industry as it provides the necessary properties required for the continuous function of the machineries without wear and tear. On the similar front, once utilized, the lubricating oil leads to a huge amount of waste oil unless recovered and re-refined. Researchers continuous work for the determination of the suitable method for the refining of the utilized lubricating oil for reuse.

In the present work, the various process used for the refining of the crude oil is discussed. Primarily, the solvent extraction process proves to be an economical and effective process where the minimal utilization of the solvent lead to the tremendous efficiency in the recovery of the lubricating oil from the used oil. The comparison and effectiveness of different solvents such as Methanol, Ethanol, Iso-propanol and others are discussed.

The evaluation leads to the judgment of the utilization of the methanol as an effective and efficient solvent for the recovery of the used oil. This has been verified by the comparative studies of the properties of the fresh lubricating oil and the re-refined lubricating oil. Even the change is observed in the re-refined lubricating oil but has been an efficient outcome of the waste.

634138 Carbon Capture Technology and Natural Gas Treatment

Abstract:

Removal of heavy hydrocarbons (C8+ HHCs) and aromatic (BTX) components from natural gas prior to liquefaction is critical for continuous LNG production. BASF has developed a new, step-change technology that allows for the combined removal of heavy hydrocarbons, aromatics, and water to cryogenic specifications in a single temperature swing adsorption unit to provide treated natural gas for liquefaction. Development of this technology required innovative bed designs to avoid pre-mature molecular sieve degradation, product development to achieve increased BTX capacity, and improved simulation tools for increased model accuracy. Several experiments were designed to prove the molecular sieve could withstand the short cycle times required for removal of HHCs and BTX. These experiments, including the rational behind the innovative bed designs, will be discussed in this paper. Finally, the process of new material development, involving Production, R&D, and Business functions, will be explained.

634137 From Discovery to Deployment of an Aromatic Transalkylation Catalyst

Abstract:

Transalkylation of toluene and C9+ aromatics to produce mixed xylenes is a key technology in modern aromatic complexes. An effective catalyst needs to balance acid function and metal function to achieve optimal performance. Challenging performance targets were set for a next generation catalyst, including increased activity, decreased ring loss and extended cycle length. In this work, we show how researchers took a zeolite discovery through development of a catalyst which was ultimately commercially deployed. The team was able to combine catalyst formulation fundamentals with high throughput testing and data driven tools to rapidly optimize a very broad formulation space. Over 120 unique catalyst formulations, spanning collectively a broad range of variables include zeolite type/combination, metal promoters, zeolite to binder ratios and methods of formulation/synthesis. The resultant catalyst lead was successfully scaled from the 100 gram lab scale to tens of thousands of pounds. In parallel, process experts put the catalyst through its paces to confirm the performance benefits and bring the catalyst to technical readiness and ultimately commercialization and deployment.

622287 Integrated Capture and Conversion of CO_2 into Materials (IC^3M); A Multi-Product Technology for Ccus

Abstract:

The capture, sequestration and utilization of CO_2 are needed to achieve the global goal of decarbonization. The United States Department of Energy (DOE) has recently prioritized utilization of CO_2 by recycling it into fuels and chemicals as a part of a broader negative-emissions strategy, though the energy and capital costs associated with capturing, purifying and compressing CO_2 are often overlooked. In this presentation, we describe a dynamic and flexible technology that couples solvent-based CO_2 capture with catalytic upgrading CO_2 captured in solution to produce low-carbon fuels and chemicals such as methane or methanol with high atom and energy efficiency. The IC^3M approach provides reduced costs while improving efficiency of detached capture and conversion processes. IC^3M allows for multiple products to be made from a single solvent process by simply changing the reagent co-feed and/or catalyst. The ability to produce multiple products allows IC^3M to change to market needs providing superior economic adaptability that single product technologies cannot achieve. We describe here, results from a multidisciplinary study that couples solvent-based capture and catalytic upgrading of CO_2 that is captured in solution to make multiple products. We provide a discussion of the energetics and economics of IC^3M and its viability in the global decarbonization market.

631547 Developing Biomass Valorization Technologies Based on Functionalization Chemistry

Abstract:

The production of various chemicals from biomass usually involves the depolymerization of biomass's three constituent polymers: cellulose and hemicellulose, which are both polysaccharides; and lignin, which is a polymer of phenyl propanoid sub-units. Achieving high yields of useful products from all three fractions is often complicated by low yields both during depolymerization and after the multi-step conversion processes that are often necessary to achieve a useful commercial biobased product. Various degradation reactions of both sugars and lignin intermediates are particular detrimental to achieving products at high yields.

Here, I will discuss how we can use functionalization chemistry, both during lignin extraction and polysaccharide depolymerization, to reversibly "trap" stabilized intermediate molecules, and facilitate their high-yield upgrading. Notably, we can show that, under the right conditions, extracted lignin can be catalytically upgraded at high yields to as few as 1-2 major products. Acetal functionalized lignin and carbohydrates could also serve as new platform molecules for the production bio-based materials including polymers. Altogether, this chemistry offers avenues for simplified scale-up and process development for the production of both high-value and bulk chemicals. I will discuss how these routes offers a realistic path towards de-risking a catalytic biorefining technology and potential commercial development.

632684 Heavy Oil Reforming in a Dual Circulating Fluidized Bed Reactor

Abstract:

TDA Research has developed a patented process to steam reform any hydrocarbon feedstock (ranging from methane to VTB) over nickel-based reforming catalysts while simultaneously regenerating the catalyst in a dual circulating fluidized bed (DCFB) reactor system. In this sense, our process functions somewhat like an autothermal chemical looping steam reformer (a-CLR), preventing deactivation of the catalyst by removing coke and sulfur in the regenerator before it can build up to damaging levels. TDA's technology, called *HyRes*, converts low-value heavy refinery feedstocks into high H_2/CO ratio syngas. Because we use a separate vessel for catalyst regeneration, limited nitrogen is introduced into the syngas, which improves the efficiency of downstream water gas shift (WGS) and hydrogen separation operations such as pressure swing adsorption (PSA).

Refineries that process heavy, sour crudes that contain metals, sulfur, and high molecular weight aromatic hydrocarbons require considerable hydrogen to upgrade these feedstocks. TDA has been granted a patent for this technology that will allow hydrogen to be produced in refineries at a cost that is considerably lower than hydrogen produced from conventional technologies (e.g. SMR) when natural gas prices are high or purchasing hydrogen from a third party.

Under a Phase II Department of Energy SBIR grant, we have successfully designed, built, and operated a bench-scale version (Figure 1) of the DCFB reactor system and have gained important knowledge relevant to the scale-up of our *HyRes* technology. We continue to experiment in our laboratories with various feedstocks, steam-to-carbon ratios, and oxygen content in the regenerator. TDA is also building a pilot-scale skid-mounted reactor system (Figure 2) to demonstrate this technology at the next scale at a refinery or chemical plant field site. Fabrication of the skid and reactor system is underway and we expect to be able to demonstrate our skid-mounted unit in 2022.

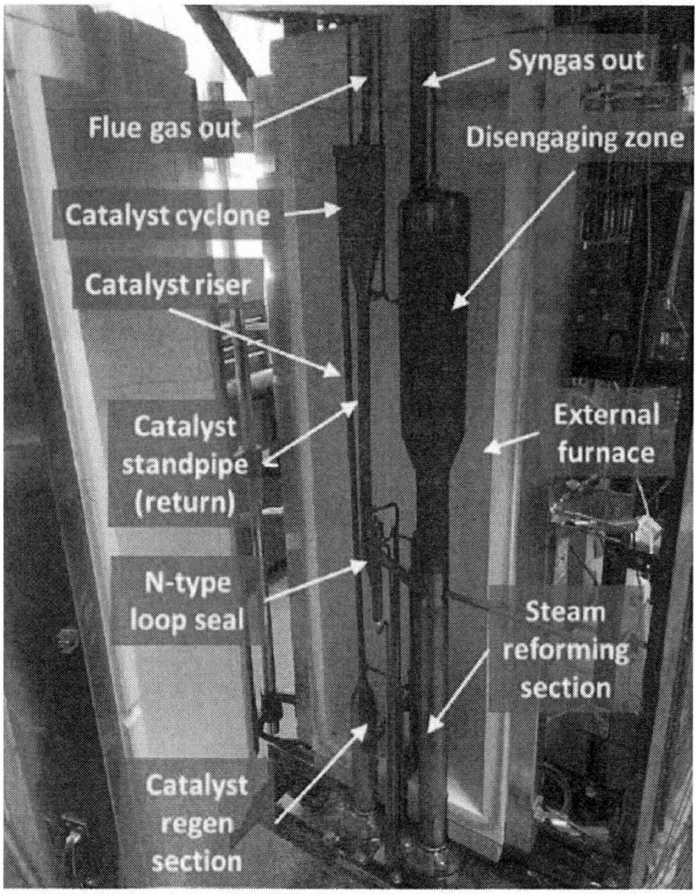

Figure 1: Bench-scale dual circulating fluidized bed (DCFB) reactor system currently operating at TDA.

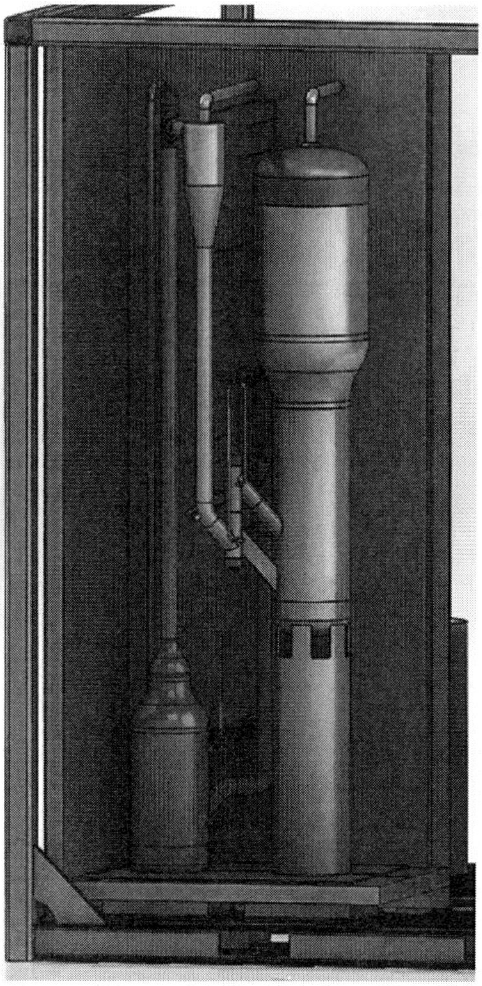

Figure 2: 3D Render of the pilot-scale DCFB reactor currently in fabrication.

629795 Catalyst Development and Process Intensification of a Bio-Renewable Surfactants Platform

Abstract:

Surfactants are the key active ingredient in cleaning products, with long lists of additional builder ingredients added to boost function while maintaining product safety and shelf life. These builder chemicals increase product cost and volume and biodegrade poorly. Sironix Renewables has invented a new class of surfactants, called Oleo-Furan Surfactants (OFS), which eliminate the need for these additional chemicals, reducing volume and resulting in a product that biodegrades readily. The OFS technology links the function of bio-based furan building blocks with natural oils to produce multifunctional and eco-friendly cleaning products. The technology gives improved performance, and by eliminating builder chemicals, reduces volume (and therefore packaging) by up to 30%. These features result in a reduced overall energy consumption while producing a more environmentally friendly product. The joint research project between Sironix Renewables and Los Alamos National Laboratory is designed to leverage the catalytic reaction engineering, catalyst development, and furan chemistry resources of the DOE national laboratories with the surfactants platform of Sironix Renewables to accelerate the DOE-invented and DOE Small Business Innovation Research-funded technology toward market commercialization. This presentation will detail efforts toward process improvements to achieve efficient scale-up of existing surfactants and the development of a new class of furan-based structures to facilitate a greater understanding of their underlying structure-function relationships.

623920 Test Tubes to Tons: De-Risking Pyran's Novel Catalytic Pathway to Bio-Based 1,5-Pentanediol

Abstract:

By employing heterogeneous catalysis, we are able to selectively remove oxygen and effectively manipulate reaction pathways to produce a suite of valuable chemicals from biomass in an economical matter. By combining expertise in heterogeneous catalysis and technoeconomic analysis, Pyran's founding team was able to identify a technically and economically advantaged pathway to produce biomass-derived 5-carbon molecules, including 1,5-pentanediol.

This talk will explore the key factors at the academic and early company stages that can lead to the identification and de-risking of novel catalytic pathways and new-to-market bio-based products. In Pyran's experience, the key to an academic project's potential to generate a commercializable technology is the simultaneous exploration of fundamental catalysis/chemistry and focus on scale-up factors and plant-scale economics throughout the program. It is also wise to take into account product and market considerations at this early stage.

Once a potentially commercializable technology has been identified, or even a startup company formed, it is necessary to consider multiple aspects for de-risking the technology and product; including determination of a cost- and time-effective process and catalyst scaleup strategy, integration of upstream (i.e. feedstock) and downstream (i.e.. polymerization) processes, identification of potential markets and customers, and strategizing early stage financing.

626054 Technology Transfer for Producing Sustainable Aviation Fuels from Wet Waste

Abstract:

With the increasing demand for low-carbon sustainable aviation fuels (SAF), new conversion technologies are needed that can transition from the laboratory to the marketplace while meeting cost and carbon intensity reduction targets. Wet waste is one such promising low-cost, widely available feedstock with the energy potential to displace over 20% of U.S. jet fuel consumption. Recent efforts at NREL have shown that wet-waste derived volatile fatty acids (VFAs) can be catalytically converted into SAF that is suitable for blending into conventional aviation fuels. This talk will examine ongoing work at NREL to partner with industry and advance VFA-SAF catalyst technology towards commercialization. Findings will be presented in regard to i) derisking and scaling VFA ketonization catalyst technology from the bench to gallon-per-day production, ii) navigating the ASTM jet fuel qualification process for new SAF routes, iii) integrating VFA-derived ketones with existing biorefinery infrastructure for hydrotreating fats, oils, and greases, and iv) quantifying VFA-SAF lifecycle carbon intensity reductions for the Low Carbon Fuel Standard. While this talk will focus on wet waste VFA-SAF, the topic of technology transfer holds relevance for advancing other emerging SAF pathways needed for decarbonization.

629162 Methane Mass Transfer in Liquid Hydrocarbon Mixtures – Measurement and Modeling in Bulk and in Porous Media

Abstract:

Methane (CH_4) dissolution and mass transport in liquid hydrocarbon mixtures is of key interest in enhanced recovery from tight shales. In this project, we have studied CH_4 dissolution and mass transport in normal alkane mixtures, both in the bulk and when confined in model nanoporous silica media at subsurface relevant conditions. The objective here is to promote a better understanding of the impact of nanoconfinement on solubility and mass transfer.

For the measurement of CH_4 dissolution/diffusion in bulk liquid hydrocarbon mixtures, we utilized a high pressure and temperature, constant-volume diffusion set-up. We studied three different normal alkanes: Decane (C_{10}), dodecane (C_{12}) and hexadecane (C_{16}). We measured CH_4 dissolution and diffusion, at formation-relevant conditions, in its binary mixtures with the three normal alkanes, as well as in ternary and quaternary mixtures. During the experiments, the swelling of the liquid mixture due to CH_4 dissolution was measured in situ via a cathetometer, and was subsequently integrated into the data analysis. A key conclusion from this part of the study is that the thermodynamic and transport properties of the multicomponent mixtures can be well predicted from the binary mixture measurements via the use of an appropriate Equation of State (EOS) and the application of the classical Stefan-Maxwell (SM) diffusion theory. This observation facilitates accurate analysis and prediction for diffusive mass transfer in more complex liquid hydrocarbon mixtures.

To measure mass transfer in the alkane mixtures under nano-confinement, we employed a high pressure and temperature Wicke-Kallenbach diffusion cell. Measurements were performed for model porous materials, prepared from silica, with a narrow pore size distribution (~4 nm). The porous disks were impregnated with the alkane(s) and placed in-between the two half-cells of the setup. The top cell is pressurized with gaseous CH_4, while the liquid (pure or mixture) flows in the bottom cell. By monitoring the pressure drop in the top cell, we can infer the rate of dissolution and diffusion of CH_4 through the model porous silica. A direct comparison of the experimental observations with the bulk phase measurements, allows us to delineate the impact of nano-confinement on diffusive mass transfer.

630355 A Liquid Heat Capacity Limit for Organic Compounds

Abstract:

The heat capacity of liquids is an important property when designing any chemical process. While there have been numerous prediction methods and quantitative-structure-property relationships developed for predicting the liquid heat capacity, all of these include some degree of uncertainty. When making capital investments, reducing uncertainty as much as possible is desirable and profitable.

In this work, a limit for heat capacity was developed as a function of temperature and the reduced temperature. This limit has been tested for hundreds of species and works for all tested organic molecules. We discuss the basis of this limit, family exceptions, and the applicability in industry and process design.

Coupling this limit with traditional methods of estimating liquid heat capacity can give a chemical engineer a higher confidence in liquid heat capacity prediction and the tools to plan safer, more cost-effective processes.

629882 Chemical Looping for Modular Conversion of Rejected Ethane to Liquid Fuels: Feasibility and Economics

Abstract:

Ethane has transport challenges for shale-producers at geographically isolated locations; it is not easily liquefied for transport, and dedicated ethane pipeline capacity is limited. This leads to frequent ethane rejection, where ethane is dumped back into natural gas for transport in pipelines. Furthermore, the concentrations of ethane in wet-gas exceeds natural gas pipeline specifications, leading to potential flaring. We have proposed a transportable, modular scale, ethane to liquids (ETL) technology enabled by low-temperature (<750 °C) chemical looping oxidative dehydrogenation of ethane (CL-ODH). In CL-ODH, a redox catalyst selectively donates lattice oxygen to convert ethane to ethylene and water. By use of automated valve switching, air is introduced to the reactor, in a separate step, to regenerate the depleted redox catalyst. The exothermic air regeneration generates heat - allowing for the system to be operated in an autothermal mode. This saves significant amounts of energy compared to traditional steam cracking, which requires high temperature furnaces (>1000 °C). The low-temperature operation of and use of parallel packed beds in CL-ODH makes it well suited for modular (and consequently lower-cost) implementation. Although ethylene is valuable, separation and transport at geographically isolated facilities is difficult. In ETL, the ethylene-rich product generated by CL-ODH is compressed and oligomerized to energy-dense, transportable liquid fuels.

The mature oligomerization technology in the proposed ETL system is well studied, but there are several technical and economic concerns to address for CL-ODH. Additional study is needed to assess the performance and lifespan of the redox catalyst, and in-depth techno-economic analysis of the CL-ODH-based ETL is needed to further increase confidence. In this work we present experimental, process modeling, and techno-economic analysis (TEA) of the CL-ODH-based ETL system. Longer-term, 500+ hr. laboratory testing demonstrates the excellent performance of the catalysts and gives high confidence in their long-term durability and activity. The ASPEN-based process modeling and TEA indicate the potential for profitable operation at typical gasoline prices. The process economics are shown to be capital driven, giving rise to nonobvious considerations for recycle. Furthermore, the conditions under which electrical cogeneration is viable are explored.

631884 Improving Hydrocarbon Yield Via Co-Feeding in Methane Coupling Reaction on Iron/Quartz Catalyst

Abstract:

Direct non-oxidative methane conversion (DNMC) over iron/silica (Fe/SiO_2) catalysts is a promising solution for developing fuels and chemicals from methane. The endothermic DNMC reaction requires high energy input to reach temperatures required for desirable methane conversion. Co-feeding oxygen (O_2) over quartz-supported Fe-species (Fe/SiO_2(Q), "Q" denotes quartz) catalysts resulted in greater than 100% increase methane conversion and 50% increase in product yield while supplying heat for DNMC from energy released during CO and CO_2 formation. Fe/SiO_2(Q) was prepared via flame-fusion, maintaining the quartz SiO_2 crystalline phase, unlike melt-fusing methods resulting in a phase transformation to cristobalite. In DNMC, the catalyst had only a brief induction period and good stability. At low concentrations, the O_2 co-feed increased methane conversion and hydrocarbon yield, suggesting that oxidative coupling of methane promotes C_2 and aromatics formation. Analyses of spent Fe/SiO_2(Q) catalysts indicated that the O_2 co-feed alters the coking species that actively participate in DNMC.

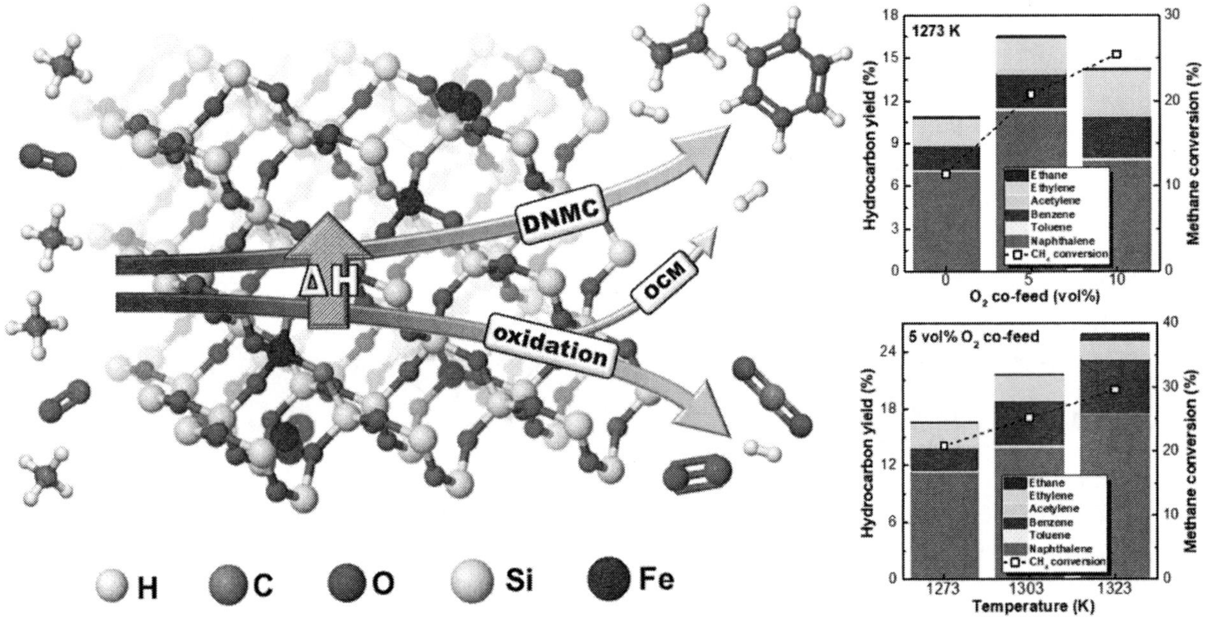

632752 Ni/Siralox Catalysts for Ethylene Oligomerization: Effects of Si/Al Ratio on Ni Speciation and Catalytic Performance

Abstract:

1. Introduction

In the United States, approximately 250 billion ft^3/yr of natural gas is flared at the production site resulting in an egregious waste of a finite resource and significant CO_2 emissions. One potential solution is on-site modular conversion of stranded gas to hydrocarbon liquids, which can be transported by tanker. Natural gas liquids, consisting primarily of ethane and propane, are the easiest target for conversion to higher hydrocarbons. The catalytic conversion of ethane to ethylene by oxidative dehydrogenation provides a feedstock for oligomerisation to C4-C14 liquid fuels, the focus of the present work. Nickel supported on Siralox, a silica-doped alumina manufactured by Sasol, has been identified as a highly efficient heterogeneous ethylene oligomerization (EO) catalyst [1]. Ni/Siralox-30 prepared by incipient wetness impregnation (with the suffix indicating the support Si content in wt.%) has received recent attention, demonstrating high conversion and good selectivity to liquid-range oligomers. The supports are manufactured by anchoring a silica precursor to boehmite so they are essentially biphasic, with the surfaces exposing 'islands' of silica or alumina (depending on which species the bulk most resembles) [2]. The effects of Si/Al ratio on Ni speciation and catalytic performance have not been investigated. Moreover, Ni/Siralox catalysts have not yet demonstrated economically viable yields of liquid hydrocarbons. Some have inferred that the active sites are isolated Ni(II) ions grafted to silanol or aluminol defects, but others have suggested coordinatively unsaturated Ni(II) ions on the surfaces of NiO nanoparticles are the active sites. Herein, we prepared Ni/Siralox-5, Ni/Siralox-30 and Ni/Siralox-70 catalysts, as well as Ni/SiO_2 and Ni/Al_2O_3 catalysts, to elucidate the influence of support composition on Ni speciation and ultimately, EO performance. The catalysts were characterized by N_2 physisorption, x-ray diffraction (XRD), diffuse reflectance infrared Fourier transform spectroscopy of adsorbed carbon monoxide (CO DRIFTS), H_2 temperature-programmed reduction (H_2 TPR), and NH_3 temperature-programmed desorption (NH_3 TPD). Initial catalyst screening was conducted in a micro-flow reactor and further testing was performed at the mini-pilot scale under industrially relevant conditions.

2. Materials and Methods

Siral-70, Siral-30 and Siral-5 silica-alumina hydrate powders and Siralox-30 and Siralox-70 silica-alumina oxide pellets were provided by Sasol. The Siral powders were calcined at 500°C for 4 h in air to convert them to the corresponding oxides. Herein, the calcined Siral powders are referred to as Siralox supports. Supported Ni catalysts were prepared by incipient wetness impregnation of the Siralox supports and pellets with aqueous solutions of $Ni(NO_3)_2$ (Sigma-Aldrich). The intended Ni loading was 5 wt.%, and impregnation was followed by calcination 500°C for 4 h in air. CO DRIFT spectra were recorded using a Bruker Vertex 70 FTIR spectrometer with Harrick Praying Mantis cell. H_2 TPR and NH_3 TPD, and BET analysis were performed with Micromeritics ASAP 2920 and 2020 instruments, respectively. XRD patterns were recorded with a Rigaku SmartLab instrument using Cu Kα radiation. ICP-AES analysis was by Galbraith Laboratories. Bench-top EO was performed in a 0.5" ID 316-SS continuous flow reactor at 225 °C under 10 bar ethylene using 0.25 g of Ni/Siralox powder catalyst (1.5 g/h ethylene, WHSV=6 h^{-1}). Products were sampled on-line and analyzed with a Shimadzu 2010 GC-FID. EO was performed also with a 1" ID 316-SS continuous flow mini-pilot scale (MPS) reactor at 130 °C under 25 bar ethylene using 15 g of pelletized Ni/Siralox catalyst (30 g/h ethylene, WHSV=2 h^{-1}). A condenser was placed downstream to trap liquid products which were periodically collected and analysed by a Shimadzu 2010 GC-FID. The off-gas flowrate was measured by a digital flowmeter and analysed on-line as previously described.

3. Results and Discussion

3.1 Catalyst characterization

The supported Ni catalysts prepared by impregnation of the Siralox supports were characterized using a variety of techniques. For Ni/Al_2O_3, Ni/Siralox-5 and Ni/Siralox-30, the XRD patterns of the Ni-containing and bare supports were indistinguishable, exhibiting broad, low intensity peaks indicative of γ-Al_2O_3. For Ni/SiO_2 and Ni/Siralox-70, the (111) and (200) diffraction peaks of NiO are visible. Those supports with a majority of amorphous SiO_2 at their surfaces favor the formation of NiO particles.

The reducibility of Ni species on Ni/Siralox catalysts is determined by the strength of interaction between the Ni precursor and the support. Bulk NiO was used as a TPR reference with a sharp peak at 400 °C (figure 1). Ni/SiO_2 reduces with a broad peak centred at 400 °C, confirming the presence of bulk-like NiO. Considering Ni/Al_2O_3, the Ni precursor is highly reactive with the alumina surface during calcination, forming a surface spinel-like $NiAl_2O_4$ phase which reduces only above 700 °C [3]. We infer that Ni/Siralox-5 contains a surface-spinel similar to Ni/Al_2O_3. Ni/Siralox-70 exhibits TPR peaks assigned to NiO reduction, $NiAl_2O_4$ reduction, and isolated Ni(II) ions grafted to acidic silanols (peak at 550 °C) [4]. Ni/Siralox-30 shows a mixture of grafted Ni(II) and $NiAl_2O_4$, although clearly skewed towards the latter. Room-temperature CO DRIFTS spectra of the Ni/Siralox powders are presented in figure 2. Ni/SiO_2, which contains bulk-like NiO particles, did not exhibit any carbonyl bands, indicating a very low dispersion. The 2184 cm^{-1} band in the spectra of Ni/Siralox-5 is the likely the result of the CO interacting with Ni(II) ions at the surface of the defective spinel-like $NiAl_2O_4$ [3]. Ni/Siralox-30 also exhibits the 2184 cm^{-1} band and a smaller component at 2200 cm^{-1}. Ni/Siralox-70 demonstrates a main band at 2200 cm^{-1} and smaller component at 2184 cm^{-1}. The 2200 cm^1 peak arises from CO interacting with isolated Ni(II) sites grafted to acidic silanols [5]. The low magnitude of the peak indicates, as suggested by TPR, that only a fraction of the Ni exists in this form.

Compared to the bare support, Ni/Siralox-5 exhibits a new NH_3 TPD peak at 200 °C that can tentatively be assigned to NH_3 interacting with Ni(II) ions on the surface of the $NiAl_2O_4$ phase, exhibiting weak Lewis acidity. In contrast, the profiles of Ni/Siralox-30 and to an even greater extent, Ni/Siralox-70 lose TPD peak area at ~250 °C on introduction of Ni. This suggests the loss of acidic silanols after grafting with isolated Ni(II) ions.

3.2 Ethylene oligomerization

The results of bench-top EO experiments at 225 °C, 10 bar, and WHSV = 6 h^{-1} for Ni/Siralox powder catalysts are displayed in figure 3, including ethylene conversion and oligomer selectivity. Ni/SiO_2 and Ni/Al_2O_3 were inactive for EO. The 25-50 nm NiO particles exhibited by Ni/SiO_2 apparently are unable or are not well enough dispersed to catalyze the EO reaction, nor are Ni(II) ions on the surface of the $NiAl_2O_4$ spinel-like phase. This explains why Ni/Siralox-5 displays only marginal EO activity. Ni/Siralox-30 and Ni/Siralox-70 are both highly active, achieving 55% and 65% steady-state conversion, respectively. Ni/Siralox-70 to our knowledge has not been reported previously as an EO catalyst. Ni/Siralox-30 and Ni/Siralox-70 contain isolated, grafted Ni(II) ions that we infer are the active sites. These catalysts displayed an approximately Schulz-Flory distribution of oligomers with high selectivity to C4 and very low selectivity to odd-numbered oligomers (reflecting the low acidity of the bare supports).

Ni/Siralox-30 and Ni/Siralox-70 pellets were tested in a MPSR at 130 °C, 25 bar and WHSV = 2 h^{-1} under more industrially relevant conditions. As shown in figure 4, both catalysts displayed high conversion (90%+) over 5 h time on stream (TOS). At an ethylene feed rate of 0.425 l/min (STP), the overall liquid yields were approximately 75%, and the liquid production rates were ~1.3 kg liquid/kg catalyst.h^{-1}. These catalysts are differentiated by their oligomer selectivity. Ni/Siralox-70 demonstrated a 40% butene mass selectivity and higher selectivity to fuel-range (C6+) oligomers, whereas Ni/Siralox-30 demonstrated 55% butene selectivity. This is reflected in their respective Schulz-Flory chain growth probabilities: 0.45 and 0.27 for Ni/Siralox-70 and Ni/Siralox-30, respectively. Characterization data suggests that this is a result of the higher concentration of isolated Ni(II) ions grafted to acidic silanol defects in Ni/Siralox-70 compared to Ni/Siralox-30, which we posit to be the EO active site. Future work will focus on EO runs over longer times on stream (24h+) to ascertain the long-term stability and liquid oligomer fuel production capacity of Ni/Siralox-70.

4. Conclusions

Ni/Siralox-30 and Ni/Siralox-70 are high-performance catalysts for EO that can be employed as part of a modular gas-to-liquid process, potentially minimizing the waste of stranded natural gas resources. Under industrially relevant conditions, they demonstrate a high yield of liquid oligomers. We infer that their performance owes to the presence of isolated Ni(II) active sites grafted to silanol defects on the support surface. In general, the activity of Ni/Siralox catalysts depends on support composition, which determines Ni speciation.

Acknowledgements

Funded by the RAPID Institute under contract number DE-EE0007888-05-6.

References

1. Lee, M.; Yoon, J. W.; Kim, Y. *et al., Appl. Catal. A-Gen.* **2018,** 562, 87-93
2. Busca, G. *Today.* **2019,** 357, 621-629
3. Garbarino, G.; Escribano, V. S.; Finocchio, E.; Busca, G. *Catal. B.* **2012,** 113-114, 281-289
4. Clause, O.; Bonneviot, L.; Che, M. *Catal.* **1992,** 138, 195-205
5. Mihaylov, M.; Hadjivanov, K. *Langmuir.* **2002,** 18, 4376-4383

Table 1 – Ni/Siralox catalyst summary

Catalyst	BET surface area (m^2/g)	BJH pore size (nm)	Total pore volume (cm^3/g) [a]	Ni loading (wt. %) [b]	Surface Si/Al ratio (mol/mol) [c]
Ni/Siralox-70	288.9	15	1.16	4.99	1.96
Siralox-70	355.8	16	1.50	n/a	2.48
Ni/Siralox-30	349.6	10	0.79	4.50	0.76
Siralox-30	515.2	10	1.01	n/a	0.73
Ni/Siralox-5	233.7	15	0.65	4.58	0.18
Siralox-5	275.0	15	0.71	n/a	0.16

[a] at P/P_0 = 0.99 [b] determined by ICP-AES [c] determined by XPS

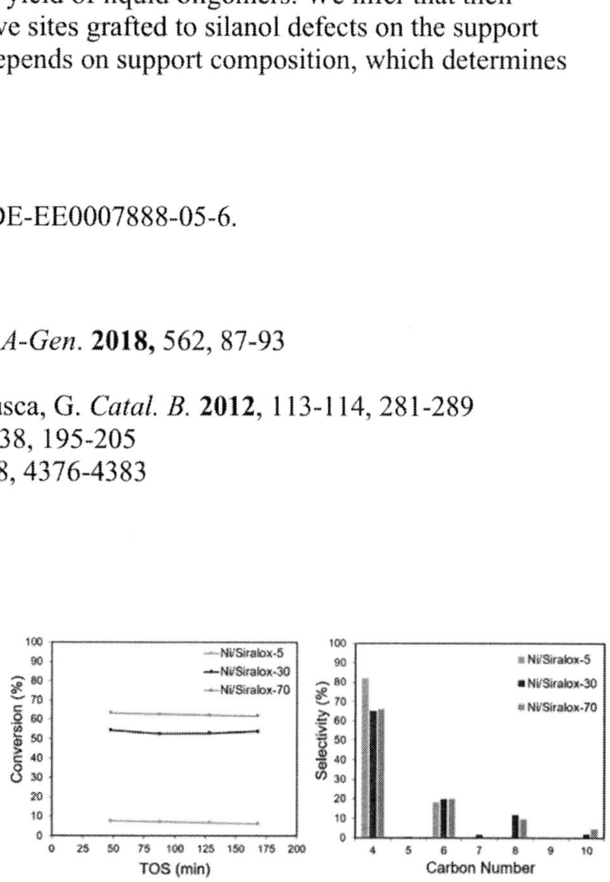

Figure 3 – Bench-top EO conversion and oligomer selectivity for Ni/Siralox catalysts

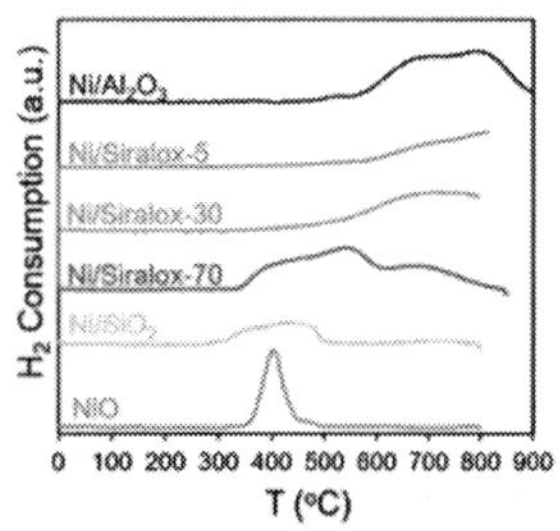

Figure 1 – H_2 TPR profiles for Ni/Siralox catalysts

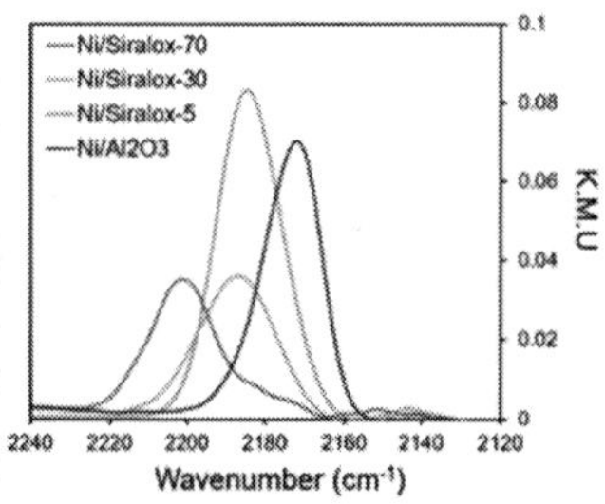

Figure 2 – DRIFT spectra of adsorbed CO for Ni/Siralox catalysts

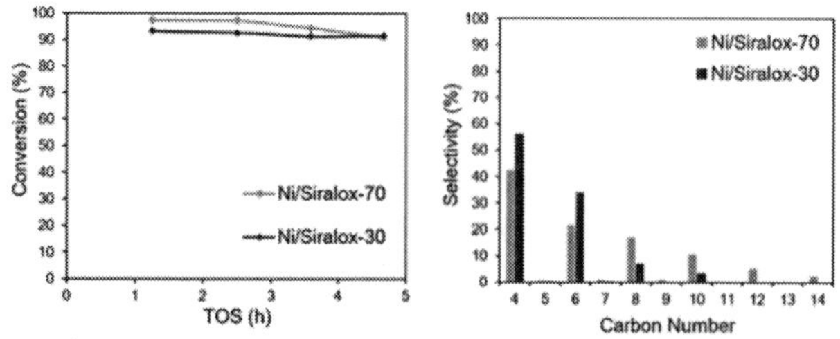

Figure 4 – MSP reactor EO conversion and oligomer selectivity for Ni/Siralox catalysts

627677 C123: From Methane to Value-Added C3 Chemicals Via Ethylene

Abstract:

Introduction

Natural gas is and will remain emerging as key player in the global energy sector. The conversion of natural gas beyond flaring, hydrogen production and capture, and technologies such as gas-to-liquids is thus critical to achieve a more sustainable chemical industry. Here a breakthrough innovation for upgrading methane to C3 commodity chemicals is presented in the H2020 C123 project[1]. Methane is firstly converted to a mixture of ethylene and synthesis gas to be further converted to C3 chemicals. The first step is denoted as the Oxidative Conversion of Methane (OCoM) while the second step is the well-known hydroformylation of ethylene to propanal which can be further converted to, e.g., propanol and propylene.

Oxidative Conversion of Methane

OCoM is a process comprising several reactions built around and critically relying on the Oxidative Coupling of Methane (OCM). While ethylene production from natural gas *via* OCM is already investigated for almost 40 years, its commercial implementation is held back due to low C2/C2+ yields, remaining below 30%, and the economic viability remains a dream. To overcome these challenges C123 aims at the conversion of methane to an equimolar mixture of ethylene and synthesis gas, rather than optimizing the ethylene yield. The mixture is subsequently converted to C3 species in the hydroformylation step, see below, ensuring a high atom efficiency in the C3 product while bypassing the challenging ethylene yield maximization.

For OCM, almost the entire periodic table has been probed as potential elements for catalytic materials. Metal oxide catalysts have shown promising results, yet the $NaWMn/SiO_2$ has recently gained attention[2] and is considered the C123 benchmark catalyst. Next to OCM additional reactions to increase the ethylene yield as well as ensure the equimolarity of the synthesis gas and ethylene mixture are considered, with post bed ethane cracking, (dry) methane reforming and (CO_2 induced) oxidative dehydrogenation of ethane to ethylene as the most promising ones. The OCoM process proposal is visualized in Figure 1. The feed is first converted to ethylene, CO and H_2, with CO_2 and water as side products, in the OCM reactor at 700-800°C. The ethylene yield is further increased in the PBC unit, where ethane is thermally cracked to ethylene. Such a configuration allows running the PBC on the heat generated by OCM and generates a stream enriched in ethylene (and H_2) at a temperature of 600°C. Finally, in the CO_2-ODH reactor, the ethylene yield is further increased while CO_2 generated during OCM, or present in a feedstock such as biomethane, is converted to CO. If required, additional synthesis gas can be generated *via* methane dry reforming, which then operates in parallel to the three reactions discussed above. To produce the most suited ethylene hydroformylation feed challenges on every scale are addressed, with fundamental modelling as the main tool to achieve the project goal.

Ethylene Hydroformylation

The produced mixture of ethylene and synthesis gas is subsequently converted to propanal *via* hydroformylation. Propanal as such is an interesting building block with applications such as, but not limited to, propanol and propylene and is furthermore easy to store and transport[3]. Currently, industrial-scale hydroformylation is a homogeneously catalyzed process because of the high ethylene conversion and even higher propanal selectivity that can be achieved using Rh complexes. Even though these catalysts are (much) more expensive than the cobalt catalysts, the catalyst selection is justified by the high propanal yield that is achieved. To reduce (precious) metal losses and facilitate product separation heterogeneously catalyzed hydroformylation is envisaged. To achieve these goals the homogeneously catalyzed ethylene hydroformylation is investigated and the most promising heterogeneous hydroformylation catalysts are designed by tethering the best performing complexes in porous supports.

Material and process design and optimization

Despite most, if not all, of the technologies considered within the project are already (well) established, the development of an integrated C123 process is not straightforward. The OCoM step is ideally operated at temperatures in the range between 650 and 900°C and at atmospheric pressure, while the hydroformylation step is ideally performed at 100°C and a total pressure between 20 and 40 bar. To ensure a smooth transition between these steps and keep all components in the gas phase, a heterogeneously catalyzed hydroformylation step is developed, see also above. To achieve these goals experimentation and (micro)kinetic modeling are performed for both the OCoM and the hydroformylation step. In the OCoM step the focus is on the effect of CO_2 and exploring the performance when operating above atmospheric pressure. In the hydroformylation step the effect of CO_2 as well as the potential to operate at a lower total pressure and a higher temperature are investigated. In this way, both steps will ultimately be brought as close as possible to each other.

The heterogenization of the hydroformylation catalyst is another challenging step in the envisaged C123 process and multiple approaches are considered, such as functionalization of silica surfaces, synthesis of a metal organic framework based material and the development of novel porous organic polymers, which can be further functionalized with, e.g. Rh. The latter is a promising heterogenization procedure that was recently discovered by Canivet et al.[4]. Following this novel synthesis method the active complex is integrated in the framework of the heterogeneous catalyst rather than attached to it. All strategies are supported by microkinetic modeling: the quantitative insights in terms of kinetic and catalytic descriptors obtained from modeling the homogeneously catalyzed hydroformylation are transferred to the heterogeneously catalyzed one, allowing for assessment the determining factors to optimize the catalyst selectivity in particular and the C123 process in general.

Conclusions and perspectives

A breakthrough technology to upgrade natural gas to propanal is presented. The process consists of OCoM, rather than OCM, where ethylene and synthesis gas are formed, and the heterogeneously catalyzed hydroformylation of ethylene to propanal. C123 is closing the gap between the operating conditions of both process steps as much as possible using a model supported multiscale approach and will ultimately develop a unified catalyst and process at TRL 5.

Acknowledgements

The C123 project has received funding from the European Union's Horizon 2020 research and innovation programme under grant agreement No 814557.

References

1. Amieiro Fonseca, A.; Heyn, R. H.; Frøseth, M.; Thybaut, J. W.; Poissonnier, J.; Meiswinkel, A.; Zander, H.-J.; Canivet, J., A Disruptive Innovation for Upgrading Methane to C3 Commodity Chemicals. *Johnson Matthey Technology Review* **2020**.
2. Gordienko, Y.; Usmanov, T.; Bychkov, V.; Lomonosov, V.; Fattakhova, Z.; Tulenin, Y.; Shashkin, D.; Sinev, M., Oxygen availability and catalytic performance of NaWMn/SiO2 mixed oxide and its components in oxidative coupling of methane. *Catalysis today* **2016,** *278*, 127-134.
3. Dubois, J.-L.; Nieder-Heitmann, M.; Letoffet, A.; Vleeming, H., C123–Methane oxidative conversion and hydroformylation to propylene. *Johnson Matthey Technology Review* **2020**.
4. Wisser, F. M.; Mohr, Y.; Quadrelli, E. A.; Canivet, J., Porous Macroligands: Materials for Heterogeneous Molecular Catalysis. *ChemCatChem* **2020,** *12* (5), 1270-1275.

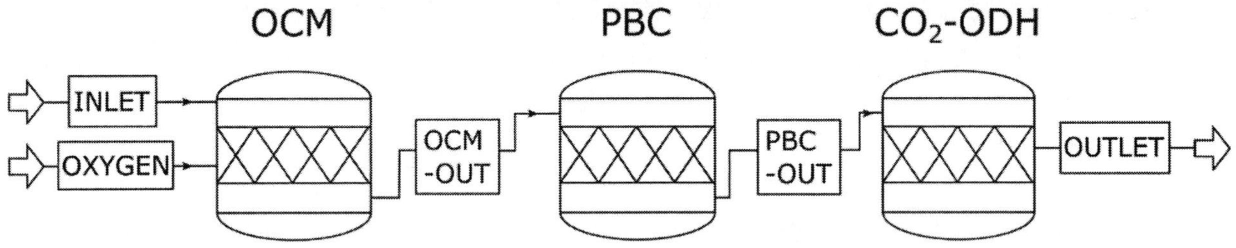
OCM
PBC
CO_2-ODH
INLET
OXYGEN
OCM
-OUT
PBC
-OUT
OUTLET

622705 Towards Elucidating and Accelerating Catalyst Activation in Methane Dehydroaromatization over Fe-ZSM-5

Abstract:

The abundance of domestic natural gas reserves continues to fuel interest in converting methane, the main component of natural gas, into higher-value chemicals. One such route is the upgrading of methane to ethylene, benzene, and hydrogen via a direct methane conversion process, methane dehydroaromatization (MDA). The main challenge of MDA lies in the methane activation step. Due to the stable tetrahedral structure of methane molecules, high temperature is thermodynamically required for the reaction to proceed efficiently, which inevitably leads to coke formation, selectivity loss, and the deactivation of the catalyst. The catalysts of choice for this reaction are metal-exchanged ZSM-5 zeolites, which require lengthy initial activation and again after each regeneration step for coke removal. Depending on the type of metal and its speciation, the induction period can take excessively long times and hence constitutes a significant hurdle toward the commercialization of MDA processes.

In the present work, the activation of Fe-based ZSM-5 catalyst was studied. Iron speciation was carefully controlled and characterized to examine its effect on the MDA induction period. We find that the induction period decreased systematically with increasing Fe loading in isomorphous substituted Fe-ZSM-5 catalysts. This suggests a correlation with an increasing amount of aggregated iron oxide clusters in the catalyst. The role of aggregated Fe species is verified by evaluating MDA on regenerated samples, where isolated Fe species aggregate to iron oxide clusters due to high local temperatures during carbon burn-off. The regenerated catalyst shows a systematic decrease in induction period with increasing regeneration cycles, confirming the acceleration of activation process by aggregated Fe species. This suggests that the presence of iron oxide nanoparticles or clusters via catalyst design could result in strongly accelerated catalyst activation. As a next step, we are now preparing Fe_2O_3@ZSM-5 catalysts via a synthetic route previously developed in our lab. In these core@shell type catalysts, iron oxide nanoparticles are encapsulated in mesopores of ZSM-5 which should enable accelerated activation of the catalyst as-synthesized. Overall, this study aims to improve our understanding of the impact of iron speciation on activation of Fe-ZSM-5 and lay the foundation for a rational design of a more efficient catalyst for MDA.

631981 Ethane Dehydroaromatization Using Molybdenum Promoted Microwave Synthesized Zeolites

Abstract:

As the demand for cleaner fuel alternatives grow, the world's use of natural gas substitutes continues to increase. Currently, natural gas makes up a large portion of the North American energy production. The conversion of natural gas to value-added chemicals involves either the direct or indirect methods. Although the indirect methods are more commonly used, the use of direct methods can be less expensive. Where indirect methods have scalable processes such as steam reforming and partial oxidation, direct methods face technical challenges. Zeolite supported metal catalysts have been extensively used in the transformation of light alkanes to aromatics. A catalyst performance and activity depend on the structure of the zeolite as well as the promotional effects of the metal promoter. Microwave energy has been recently used as a favorable way to help facilitate some chemical reactions. Microwave energy has shown promise as an energy efficient and time saving experimental method which can help improve many reaction systems. The use of microwave radiation can help speed up the time-consuming synthesis process associated with the synthesis of zeolites and potentially increase the yields compared to conventional methods. In this study, a microwave synthesizer was employed to synthesize the zeolite support, ZSM-5. The synthesized zeolite was then characterized by XRD, TEM, SEM, and BET prior to reaction testing. Following the characterization of the zeolite crystal the support was impregnated with a molybdenum promoter by incipient wetness impregnation. The prepared metal-loaded catalyst was then tested for the ethane DHA reaction in a fixed bed reactor at 650°C for two hours. The ethane conversion, product selectivity, and production rates were tracked and plotted. The spent catalyst then underwent characterization testing to determine full catalytic reaction performance during the ethane dehydroaromatization reaction.

624210 Feasibility of Chemical Looping Ammonia Synthesis from Metal Nitrides and Hydrides and Their Alloys.

Abstract:

Ammonia is a commercially important chemical to the food, drug and energy industry. In 2015, 160 million tons of synthetic ammonia was produced to meet global demand which is expected to rise 1.5% yearly [1]. The Haber Bosch process is a century old process, responsible for the majority of industrial ammonia synthesis through the direct combination of its elements: $N_2 + 3H_2 \leftrightarrow 2NH_3$ (T=450°C, P=200atm). The Haber Bosch process is an energy intensive polluting process, accounting for 1% of the world's annual energy demand while emitting 400 Mt/yr of CO_2 [2]. The significant negative environmental impact of the Haber Bosch process has led to a search for alternative means to synthesize ammonia [2]. Chemical looping ammonia synthesis (CLAS) has been proposed as a low energy alternative to the Haber Bosch process [3]. Through a multistep process, CLAS circumvents the equilibrium of the Haber Bosch process and its high pressure requirement [3]. CLAS processes mediate ammonia synthesis by fixing nitrogen to a compound, then reacting that compound with a hydrogen carrier to yield ammonia, e.g., 1) $3MH_2 + N_2 \leftrightarrow M_3N_2 + H_2$, 2) $M_3N_2 + 6H_2 \leftrightarrow 3MH_2 + 2NH_3$ (T=100°C, P=1atm), where M is the material (hydrogen or nitrogen) carrier. Identifying carriers M, capable of mediating ammonia synthesis in a chemical loop is difficult, due to the excessive number possible material-reaction combinations, the majority of which are infeasible [3]. Additionally, tradeoffs exist between the ability of an element to fix nitrogen and then release it as ammonia in the presence of a hydrogen donor [3].

Prior analysis has identified a small number of CLAS schemes that are theoretically capable of producing ammonia at appreciable amounts without the energetically expensive requirements for high pressure of the original Haber Bosch process. Limited experimental evidence [3] supports the hypothesis that material mixtures (usually in form of metal alloys) offer advantages in the form of bifunctional carriers that allow for increased ammonia yields. To identify metals and their alloys that maximize the theoretical yields of ammonia via low-pressure, multi-step CLAS schemes, we present a comprehensive framework for evaluation of schemes and material carriers along with the optimization of reactor conditions for these schemes at steady state. First, elements from a chemical database are paired with CLAS schemes of a recent literature review [4] and are evaluated for their thermodynamic feasibility and favorability. The feasible set is then expanded to alloy carriers. Successful CLAS processes are optimized in Aspen Plus, wherein the reaction steps are assumed to be governed by Gibbs free energy minimization constraints of the chemical and phase equilibrium. Then, the performance of optimized CLAS processes is compared and ranked against metrics that measure the energy demand, initial cost and environmental impact of the process. It is shown that combining a successful nitrogen fixing element with a successful hydrogen carrying element can increase the ammonia yield of a CLAS scheme. Additionally, the combination of elements allows for the use of low cost environmentally safe materials, which can accelerate the development of CLAS processes. This work concludes with a list of successful combinations of elements and chemical loops for ammonia synthesis at low pressure.

Acknowledgements
This work was partially sponsored by the United Technologies Corporation Institute for Advanced Systems Engineering (UTC-IASE) of the University of Connecticut. Any opinions expressed herein are those of the authors and do not represent those of the sponsor.

References

[1] Fao, *World fertilizer trends and outlook to 2018;* Annual Report 14; Food and Agriculture Organization of the United Nations (FAO): Rome, Italy, 2015; ISBN 978-92-5-108692-6.

[2] L. Wang *et al.*, "Greening Ammonia toward the Solar Ammonia Refinery," *Joule*, vol. 2, no. 6, pp. 1055–1074, 2018.

[3] W. Gao *et al.*, "Production of ammonia via a chemical looping process based on metal imides as nitrogen carriers," *Nat. Energy*, vol. 3, no. 12, pp. 1067–1075, Dec. 2018.

[4] L. Burrows *et al.*, (In Review) "Feasibility of Chemical Looping Ammonia Synthesis from Metal Nitrides and Hydrides and their Alloys," *Chem. Eng. J.*

624147 Hybrid Gas and Chemical Stimulation for Shale Enhanced Oil Recovery

Abstract:

Background: Production of oil from tight oil shales has more than doubled the US oil production in the last decade. Oil is produced by depressurization using horizontal wells with multiple fractures. However, extremely low permeability (~0.1 micro Darcy) and microfracture closing reduce the oil productivity by more than 70% in the first year. Depressurization produces less than 10% of the oil present in these shale oil formations. Novel technology is needed to extract more oil out of these shale formations. Treating these formations with gases and surfactant solutions are being investigated. Injection of CO2 or hydrocarbon gases can swell the oil and reduces oil viscosity; the decrease in pressure following the injection can nucleate gas bubbles in situ, push oil into the fractures and prolong the oil recovery. Such processes are called huff-n-puff processes. Injection of surfactants can change the wettability, lower interfacial tension and improve oil flow. The goal of this work is to study mechanisms of improved oil recovery due to combined gas and chemical injection and compare the effectiveness hydrocarbon gases with CO_2in huff-n-puff EOR processes for a shale oil.

Methods: Gas and chemical huff-n-puff experiments were conducted at the laboratory scale using outcrop shale core plugs. Core plugs were saturated with a light oil at a high temperature and high pressure (7000 psi) for two weeks. Initial liquid volumes and added oil volumes were evaluated using NMR T2 relaxation times. The saturated shale cores were placed in a high-pressure reaction cell to conduct experiments. All huff-n-puff experiments were conducted at the reservoir temperature (75 °C) and elevated pressure (huff: 2800~5200 psi, puff: ~2000 psi). The oil/water volume change was evaluated by NMR after the experiment to calculate the oil recovery factor. Three sets of huff-n-puff experiments were conducted, one using CO_2, another using hydrocarbon gases (C1, C2, C3 mixtures) and the third using CO2 and surfactant solutions at several pressures. The composition of hydrocarbon gas varies with reservoirs. For this work, a mixture of 60% methane, 35% ethane, and 5% propane was used. The cores were pressurized with the gas to certain huff pressure, soaked for 2 days, then depressurized to 2000 psi. In hybrid experiments, a surfactant solution was used in some of the pressure cycles along with the gas; D2O was used in place of water for NMR detection. Three huff-n-puff cycles were conducted for each experiment.

Results: NMR experiments showed that the injected oil responds at the 10 ms to 100 ms T2 relaxation time region. For a 1.5 in x 1.5 in shale outcrop, around 0.8 ml ~ 1 ml oil was injected. CO_2huff-n-puff is an efficient technique to improve recovery from oil shales. Most of the added oil was recovered in all the cases. The liquid recovery was about 40% of the fluid in place. The pressure to which the cores were pressurized with CO_2 did not affect the oil recovery significantly as long as the pressure was high enough (2800 psi). The "huff" pressures used in this study were higher than the CO_2MMP. In the hydrocarbon gas huff-n-puff experiment, the total recovered oil depended on the huff pressures. At high pressures (4000 psi and 5200 psi), the oil recovery factor was similar to the CO_2cases. But at low huff pressure (2800 psi), the oil recovery was substantially lower. Chemical blend plus $CO_2$was more effective than pure CO_2huff-n-puff (40% vs. 64%) mainly by increasing the oil recovery from the original liquid inside the shale. The surfactant can lower the interfacial tension and oil-wettability while CO_2dissolves in oil and swells the oil. The sulfate ions can delay the acid reaction allowing acid molecules to reach far into the matrix and dissolve carbonate minerals. This process can improve the stimulated volume around fractures. Pumping chemical blend to pressurize previously injected CO_2gave equally high recovery as injecting CO_2after injection of the chemical blend. This novel technique would avoid the use of large compressors.

Implications: This work evaluates using of CO2 and produced gas in addition to aqueous chemicals in Huff-n-Puff process. The use of produced gas reduces the issue of gas availability. The use of produced gas can also reduce the need for gas flare and reduce the associated environmental impact. The use of aqueous chemicals improves gas huff-n-puff. These processes are effective in the laboratory scale, but their scale-up to field scale needs to be studied.

624534 Improving CO_2-EOR in Shale Reservoirs Using Dilute Concentrations of Wettability-Altering Nonionic Surfactants during Lab-Scale Huff-and-Puff

Abstract:

The aim of this work is to determine if the efficacy of CO_2-based enhanced oil recovery (EOR) for unconventional reservoirs, which is based on numerous mechanisms (e.g. diffusion, oil swelling, oil viscosity reduction, pressure maintenance, etc.), can be improved with dilute concentrations of CO_2-soluble wettability-altering additives. Oil recovery during water-based hydraulic fracturing and water-based EOR in unconventional liquid reservoirs (ULRs) can improve when ionic and nonionic surfactants that greatly enhance the water-wetness of rock surfaces are added to the water. In this study, we assess whether surfactant-based wettability alteration can increase the amount of incremental oil recovered during the CO_2 huff and puff process. Although most ionic surfactants are essentially CO_2-insoluble, there are numerous inexpensive oil-soluble and water-soluble nonionic surfactants that are CO_2-soluble to ~0.1 – 1.0% at formation conditions. This work focuses on altering the wettability of unconventional rock away from oil-wet using these non-ionic surfactants, and then assessing whether the additional amount of oil recovered during CO_2 huff and puff is significant considering the cost of the surfactant addition.

The solubility of various nonionic surfactants (Indorama SURFONIC® N-100, an ethoxylated nonylphenyl alcohol with 10 ethylene oxide (EO) groups; Indorama SURFONIC® TDA-9, -6 and -3, ethoxylated branched tridecyl alcohols with 9, 6 or 3 EO groups, respectively; Indorama SURFONIC® TDA-8PO, a tridecyl propoxylated alcohol with 8 propylene oxide groups; and Indorama SURFONIC® L12-6, an ethoxylated normal C10-C12 natural alcohol with 6 EO groups) in CO_2 has been measured between 25 – 100 °C. The surfactants exhibit a solubility of roughly 1 wt% at pressures of 2000 - 5000 psia, with lower pressures required for lower temperatures, lower surfactant concentrations, and increased oil solubility of the surfactant.

The rock samples were first aged by via immersion in dead Eagle Ford crude oil at high temperatures and high pressure for a prolonged period of time. Then, the oil-wetness was verified by measuring the contact angle of water droplets on the shale exposed to air. To determine CO_2-induced or CO_2-surfactant solution-induced changes in wettability, the oil-wet, oil-saturated shale disk was immersed in CO_2 or a CO_2-surfactant solution for 16h at 4000 psi and 80°C. No significant changes in contact angles were observed in samples immersed in CO_2, whereas immersion of the shale in CO_2-surfactant solutions showed dramatic changes in wettability toward water-wet for some of the surfactants. Small cylindrical shale samples were then used in huff and puff tests to assess oil recovery. Ultimate oil recovery after 8 huff and puff cycles increased substantially when some of the surfactants were added to the CO_2, especially during the early cycles. A comparison of the performance of these surfactants will be presented.

625833 High-Pressure Solutions of C6F13- and C4F9- Based Polyfluoroacrylates in CO2: Solubility, Viscosity, Adsorption, and Sealing of Cement Cracks

Abstract:

Polyfluoroacrylates (PFAs) have drawn interest in recent decades due to their remarkable solubility in liquid and supercritical carbon dioxide. PFAs have been considered in many different applications in CO_2-related chemical engineering technologies and petroleum engineering projects (i.e. designing CO_2-soluble compounds and thickening CO_2 for more uniform flow of CO_2 through porous media during enhanced oil recovery). This project focuses on the use of CO_2-PFA solutions to improve wellbore integrity. Our initial attempts to seal small cracks in cement using CO_2-PFA solutions suggest that PFAs dissolved in CO_2 will strongly and adsorb onto cement surfaces. This deposition of a hydrophobic, oleophobic amorphous polymer dramatically changes wettability and reduces the gap size of the crack, both of which reducing crack permeability and – in some cases – completely sealing the crack. To better understand this crack-sealing process, we used a high pressure, high temperature, windowed, stirred, variable-volume cell to obtain PFA solubility in CO_2 and PFA-CO_2 solution viscosity for PFAs based on C6F13 and C4F9 pendant groups. We have also presented (for the first time, to the best of our knowledge) high pressure equilibrium polymer adsorption data for calcium carbonate particles immersed in liquid or supercritical solutions of CO_2 and PFA. By measuring changes in solid particle mass before and after 24 hours of immersion in PFA-CO_2 solutions of known initial PFA concentration, we determined the equilibrium amount of polymer adsorbed on solid particles of known surface area. The corresponding equilibrium concentration of PFA in CO_2 was calculated by material balance.

626708 Critical Assessment of Brittleness Indices As Proxies for Rock Mechanical Impacts on Hydrocarbon Recovery: Case Study from the Caney Shale, Oklahoma

Abstract:

In the drive to maximize gas recovery from horizontal wells, one common approach is to assign so-called "Brittleness Index" values to the rocks. The underlying premise is more brittle rocks will be stimulated more effectively by hydraulic fracturing and will be less prone to the creep related closure of fractures which leads to production decline. There are several manifestations to calculate brittleness. One approach is to characterize a brittleness index along a horizontal well, and target simulation in the most brittle sections of the well. Moving back one step in the decision-making process, a second approach makes use of the brittleness index of layers within a stratified reservoir so the landing depth of the horizontal well targets the most brittle layer(s). A third approach views the entire formation, i.e. an entire shale play, as being more or less difficult to develop based on whether it is a "brittle" or "ductile" shale.

While these approaches have some merit and can point to local successes, criticisms stem from the multiplicity of definitions, none of which comprehensively summarize rock behavior relevant to hydrocarbon production and some of which can contradict one another. At a deeper level, the language of "brittle" versus "ductile" describes shale according to two misleading end member labels, when in fact shale is a class of rocks which always fractures in a quasi-brittle manner and never truly brittle (like glass) or ductile (like clay).

This talk will focus on a three-fold critical comparison of brittleness index (BI) as applied to hydrocarbon-bearing shales. The first compares various measures of BI, ranging from evaluation of non-linearity in stress strain curves under triaxial compression testing, to comparisons of rock strengths and/or stiffnesses, to purely mineralogy-based descriptions. Using recent experimental results from a current project in the emerging Caney Shale, Oklahoma, we show the mechanical and mineralogical descriptions of BI are weakly correlated, at best, and in some cases are anti-correlated, pointing to contradiction.

A second comparison evaluates possible connection between brittleness indices and creep properties. Initial laboratory results demonstrate predictions based on brittleness index do not consistently match behavior from triaxial mechanical properties testing. However, creep testing shows nominally ductile zones are considerably more prone to viscoelastic and viscoplastic deformation compared to nominally brittle zones.

Finally, a third comparison examines brittleness indices among different plays, including nominally "brittle" plays (e.g. Barnett) and nominally "ductile" plays (e.g. Haynesville). While difficulty stems from the multiplicity of ways to define brittleness, it is possible to apply multiple brittleness definitions to each formation. Such a comparison gives rise to consensus definition of brittleness and, in this regard, Caney appears to be relatively brittle by all definitions except for the most industry's most commonly-applied definition based on elastic properties.

Hence, the distinction of "ductile" does not appear to directly apply to the Caney. However, in its vernacular use in regards to how prone a rock is to creep, the Caney does show propensity to time-dependent deformation which could pose a challenge to be addressed during the development of this emerging play.

634101 An Optimal Production/Injection Scheduling for Gas Condensate Reservoirs with Gas Recycling

Abstract:

The recycling process is a common industrial technique to improve condensate production from natural gas condensate reservoirs. The life cycle net present value (NPV) is an economic measure involving both time and cost, to evaluate the recycling process with respect to the injection and production costs.

This work highlights necessary elements and workflow for recycling design prior to implementation for improved production strategy. We will present an operational schedule for gas recycling using model-based stochastic optimization. Since life-cycle model-based optimization of a gas condensate reservoir is a time-consuming and CPU intensive task, this work suggests optimizing the process using the full compositional model of the reservoir and a simplified compositional interacting-tanks (CIT) model for calculation of objective function. In this approach, the reservoir grids are divided into two large columns of grid blocks (tanks). One of these columns contains a group of producing wells, and in the other column, the produced dry gas from the separator is recycled back into the reservoir. CIT increases the simulation speed as well as optimization in the early iterations.

After a careful analysis of three different evolutionary algorithms, a hybrid particle swarm optimization and genetic algorithm (PSO-GA) is used to maximize life-cycle NPV. To ensure applicability of the proposed method regarding the speed of optimization process and accuracy of the results, the scheduling technique is evaluated with a benchmarking synthetic model from the literature. PSO-GA optimization combined with the full/CIT model provides accurate results compared with the optimization using the full model while increasing the speed and efficiency of the optimization process. In addition, the proposed technique is used to develop a production plan for a field sector. The optimized schedule considerably increases the NPV of the recycling process.

Since there is a difference in the price of injected gas in cold and warm seasons, the optimized reinjected gas in the cold season is about half of the gas in the warm season. This study also shows the sensitivity of the optimized gas-recycling scenario to the richness of gas condensate reservoirs.

630611 A Study of Condensation Phenomena in Shales and in Synthetic Model Materials

Abstract:

The presence of gas in Shale nanopores, caused by capillary condensation, strongly affects the estimation of the quantity of hydrocarbons in place. Capillary condensation also affects the mass transport in Shales due to the low mobility of the condensed hydrocarbon. The focus of this study is on the phase behavior of hydrocarbons when confined in nanopores, since the thermodynamic fluid behavior under confinement is known to be different from that of the bulk phase. In particular, we investigate low molecular weight (MW) alkanes relevant to shale gas production, and study capillary condensation, which is the vapor to liquid phase change that occurs in nanopores below the saturation pressure of the bulk fluid. We investigate, in particular, the impact on capillary condensation of the pore size, as well as of the pressure/temperature conditions of the bulk fluid in contact with the nanopore.

For the study, we have synthesized model porous media, specifically mesoporous silica disks with narrow pore size distributions, with varying average pore size <10 nm. We have selected ethane as a model low MW hydrocarbon component of shale gas. We utilize a high-pressure thermogravimetric microbalance to study adsorption and phase condensation of ethane in the silica nanopores. The experiments are complemented with models of the phase behavior, including the classical Kelvin Equation (KE), a reformulated KE model based on the Peng-Robinson Equation of State (EOS), as well as direct molecular simulations of ethane adsorption in the silica nanopores. Models and experiments are in satisfactory agreement with each other. The next stage of our work includes experimenting with the natural Shale materials to compare their behavior with that of the model analogs.

628720 Geochemical Impacts on Mineral Scaling in Unconventional Shales: Interplay between Mineralogy, Additives, and Base Fluids

Abstract:

Unconventional oil/gas shale production has led to significant reductions in CO_2 emissions over the last decade while also providing US energy independence. Though advances in hydraulic fracturing have resulted in extraction of resources previously thought unobtainable, this stimulation technique leaves significant resources underground. Part of this is due to the low porosity/permeability of the rock, and mineral scaling caused by injection fluids. Contrary to conventional oil/gas extraction, unconventional extraction relies on high-pressure injection of fluids (acid, additives, and base fluids) and proppants to fracture rock, ease extraction of hydrocarbons, and maintain apertures of newly formed fractures. Additionally, most wells have shut-in periods between the end of injection and start of production that can last weeks to months. The injection and shut-in of fluids results in a wide array of chemical reactions to the rock that are both beneficial and deleterious. The increase of porosity/permeability due to acid injection at the beginning of a stimulation is beneficial, but can lead to deleterious mineral scaling as solution conditions change.

We have adopted a multi-faceted approach to identifying types of mineral scaling, its causes, and its effects on the shales caused by stimulation chemicals. Our approach includes batch reactions using ground and intact core material, synchrotron-based spectroscopy/imaging, reactive transport modeling, core flood experiments with contrast CT imaging, and acoustic monitoring of un/reacted samples. By using these complementary *in situ/ex situ* techniques, a clearer picture of what is occurring downhole in the stimulated rock volume is obtained.

Batch reactors for two major shale plays, Marcellus and Permian Basin, were conducted using shale and fracture fluid recipes from each respective geological area. Time-resolved and pressurized experiments were conducted using both ground shale (150-250 μm) and intact cores (1 cm x 1 cm). Experiments were conducted following a standard fluid injection sequence and proper mixing ratios at 80°C for 3-weeks to mimic field practices. Post-experiment analyses using synchrotron-based techniques show that pyrite contained in the shale partially dissolved and the released Fe(II) oxidized resulting in the precipitation of a new mineral scale in the form of Fe(III)-(hydr)oxides. The types of Fe(III)-bearing particles formed varies depending on the carbonate concentration (i.e. buffering capacity) of the shale. These experiments were conducted using doubly de-ionized (DDI) water. Another set of experiments conducted using Permian Basin samples and water from the field, both municipal freshwater and clean brine, showed significant sulfate mineral precipitation on clay-rich samples as well as Fe(III)-bearing mineral precipitates when clean brine was used. Our results show that mineral scaling occurs on both the fracture surface and within the shale matrix caused not only by alterations to the matrix but also by reactions between the injection fluid additives and the shale.

We leverage these observations to inform forward, predictive models merging the principles of transport and transformation. Such reactive transport software capabilities have been developed largely in application to contaminant hydrology, but the same principles are readily adapted to embed process understanding of mineral scaling into models encapsulating the open, transient conditions of unconventional shale stimulation. This parlay of molecular scale information into environmentally relevant context via forward models offers a means of describing how a variety of mineral scaling reactions combine with the unique condition of each reservoir to impact reactive alteration of the shale, the attendant fluid transport properties and ultimately production. Thus far, we have built a suite of multi-component reactive transport simulations using the CrunchFlow software which encapsulate the mineralogy of Marcellus and Permian shales, and the principle fluid additives to reproduce the temporal evolution of reactive fronts developing in the shale as constrained by our synchrotron work. The models demonstrate that the solubilization of bitumen is vital to convert Fe(II) released via oxidative dissolution of sulfides to Fe(III) and thus acts as a mediator of the extent to which iron mineral scaling occurs. Sulfate minerals, including barite and celestite, have also been built into these forward model simulations, and suggest a close relationship between the depth of reactive fluid penetration, the content of easily solubilized carbonates, and the accumulation of this class of mineral scaling. Such predictive capability offers a rapid diagnostic of the effects of a given fracturing fluid composition on a given shale play and ultimately offers the potential to reduce the need for re-stimulation and thus the volume of water consumed during the production life cycle.

Reactive core flood experiments are the logical extension of batch reaction tests and provide data for matching of reactive transport models. Reactive core floods are, by their nature, long experiments as it takes time to inject a suitable amount of fluid into the low permeability system. We have examined clay-rich and carbonate-rich samples from the Marcellus shale. Samples have a distinct microcrack and matrix network. Alterations to shale were measured using computed tomography of samples before, during, and after experiments. Scanning electron microscopy was used for detailed views of the shale fabric before and after exposure to reactive fluid. Samples tested to date show a reduction in gas accessible porosity following reaction. A corresponding reduction in liquid permeability was measured. SEM-EDS (energy dispersive spectroscopy) of rock surfaces shows an iron-containing precipitate on and near fracture surfaces and in the shale matrix of the clay-rich sample indicating partial dissolution of pyrite and/or ferruginous dolomite followed by precipitation of iron (hydro)oxide. Extensive imaging reveals fracture filling with migrated and/or precipitated fine particles. Significant barite scale growth was detected on the reacted surfaces of the carbonate-rich sample. Halite and (hydro)oxide precipitates resulting from geochemical reactions between the basin-specific injectant and shale minerals were also noted. The carbonate-rich sample experienced substantial calcite dissolution and a corresponding decrease in its bulk density and microcrack openings. That is, an altered zone was apparent over the roughly first ½ of the length of the core. It is clear that fracture fluid composition should be optimized based on intrinsic shale and resident brine chemistries.

Shale alterations not only result in new mineral scaling, but result in mechanical alterations to fractures and rock matrix, such as porosity/permeability increases, mineral scale precipitation, and fracture closure due to proppant embedment. To monitor these changes to the stimulated rock volume *in situ*, it is necessary to develop seismic signatures in the laboratory that benefit the interpretation of field measurement changes in velocity. We conducted time- lapse rock physics experiments on clay-rich (carbonate-poor) Marcellus shales to characterize the acoustic velocity and permeability responses to fracture acidizing and propping. We measured acoustic P- and S-wave velocities together with permeability before and after inducing chemical alterations of the sample fractures. We also complemented the physical measurements with microstructural imaging using X-ray μCT and SEM techniques. Experiments show that S-wave velocity is a key geophysical observable, and in particular, the S-wave polarized perpendicular to fractures as it is sensitive to fracture stiffness. Results show that both acidification and propping decrease the fracture elastic stiffness. This effect is stronger when fractures are acidized making the monitoring of proppant efficiency be potentially masked by chemical alteration. Furthermore, fracture permeability is compromised by the softening of fracture surfaces due to acidification. Conversely, the problem is minimized when fracture propping is highly efficient (i.e., majority of the fractures are propped). These contrasting effects on fluid flow in combination with similar seismic attributes indicate the importance of experiments on basin-specific shales so as to improve rock physics models that incorporate the coupled effects of hydraulic fracturing and acidification. This is necessary for a correct interpretation of seismic velocity signatures of flow pathways in unconventional shales.

630520 Chemical Reactivity of Caney Shale to KCl-Brines at Elevated Temperature and Pressure

Abstract:

Development of unconventional resources beginning in the past decade has created a paradigm shift in the definition of reservoir and seal rocks. Formations that were previously considered seals are now being reappraised with the possibility of being considered reservoirs. The Caney Shale located in South-Central Oklahoma Oil Province (SCOOP) is one such plays that has been appraised and is currently being developed. Due to its previous classification as seal rock, little work has been conducted on the formation's response to drilling and fracturing fluids.

Published literature on low permeability shale rock's interaction with drilling and fracturing fluids have confirmed nano-scale, micro-scale and macro-scale physical-chemical interactions with implications on engineering technologies, from drilling to production. For instance, during rock-fluid interactions in the subsurface, dissolution of carbonates create secondary permeability whilst clay swelling, dislodgement and transport of clay fines lead to significant permeability reductions. Permeability-enhancing reactions as well as permeability-reducing reactions are significant considering the magnitudes of scale of permeabilities in shale reservoirs. These reactions and outcomes, though similar for different shale reservoirs, are unique for every formation in terms of the extent of damage or enhancement. These differences are due to mineralogical and microstructural differences in various shale rocks, which are result of depositional environmental and geological history.

In this study, rock-fluid interactions between the Caney Shale and simplified fracturing fluid at *in-situ* T, P conditions were examined to ascertain the responses of the formation to these fluids. Mineralogical and microstructural analyses of Caney Shale samples were done using X-Ray Diffraction (XRD) and Scanning Electron Microscopy (SEM) methods. Results of mineralogical microstructural analyses show samples are mainly quartz, clay and carbonates with different degree of micro-fractures interspersed between the interlocking mineral grains. Core flooding experiments were conducted on propant filled fracture, simulated by two parallel plates (following API-19D), at constant temperature of 100°C to mimic formation temperatures of the Caney Shale. Confining pressures were varied from 1000psi to about 12000psi. During the experiments, effluents are collected at predetermined time intervals and analyses conducted to ascertain the elemental concentrations. Preliminary results from effluent analyses shows the presence of Ca, Si, Al, Na, B, K as well as SO_4^{-2}. These changes will be discussed in relation to fracture permeability and ultimate production of hydrocarbons from Caney Shale, Oklahoma.

Figure 1. Formation core cross section. Number of platens generated depends on dimensions and integrity of drill core. "A" and "B" denote top and bottom face of each platen in this orientation. Fracture face for conductivity testing depends on customer preference.

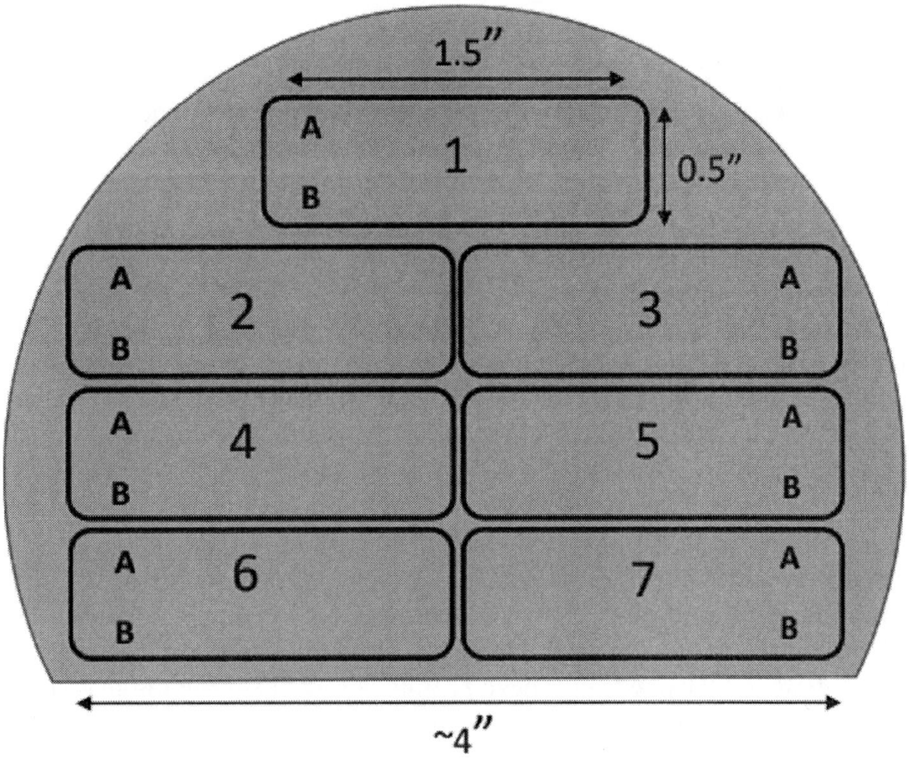

630224 Identify Fracture-Hit Events in Low-Frequency DAS Strain Rate with Convolutional Neural Networks

Abstract:

Summary. Distributed acoustic sensing (DAS) has been used in the oil and gas industry as an advanced technology for surveillance and diagnostics. Operators use DAS to monitor hydraulic fracturing activities, to examine well stimulation efficacy, and to estimate complex fracture system geometries. Particularly, low-frequency DAS can detect geomechanical events such as fracture-hits as hydraulic fractures propagate and create strain rate variations in the rock. Analysis of DAS data today is mostly done post-job and subject to interpretation methods. However, the continuous and dense data stream generated live by DAS poses the opportunity for more efficient and accurate real-time data-driven analysis. The objective of this study is to develop a machine learning-based workflow that can identify and locate fracture-hit events in simulated strain rate responses correlated with low-frequency DAS data. In this paper, "fracture-hit" refers to a hydraulic fracture originated from a stimulated well intersecting an offset well. We start with building a single fracture propagation model to produce strain rate patterns observed at a hypothetical monitoring well. This model is used to generate two sets of strain rate responses with one set containing fracture-hit events. The labeled synthetic data are then used to train a custom convolutional neural network (CNN) model for identifying the presence of fracture-hit events. The same model is trained again for locating the event with the output layer of the model replaced with linear units. We achieved near-perfect predictions for both event classification and localization. These promising results prove the feasibility of using CNN for real-time event detection from fiber optic sensing data. Additionally, we use edge detection techniques to recognize fracture-hit event patterns in strain rate images. The fracture-hit location can be identified using recognized pixels in the image. The accuracy of edge detection-based location identification is also plausible, but edge detection is dependent the assumption of pattern shape and image quality, hence less robust compared to CNN models. This comparison further supports the need for

Event Classification. CNN applications in image-based real-time fiber optic sensing event detection. A propagating fracture exhibits a characteristic cone-shaped extension zone surrounded by compression at the fracture tip (Tang and Zhu, 2021). We adopt a fracture propagation model to simulate fracture geometry and stress field around the fracture. With this approach, we generate the strain rate responses for various completion scenarios. The responses are passed to the CNN models for event classification and localization.

When installed outside of the casing of a monitoring well, the fiber optic cable is considered mechanically coupled with the casing and the formation. When a hydraulic fracture perpendicular to the monitoring well approaches the well, it causes axial deformation of the fiber. Low-frequency DAS can detect such deformation because the phase change in the backscattered light is linearly correlated with the axial displacement. Hence the displacement can be used to approximate the axial strain rate. (Jin and Roy, 2017).

Convolutional neural network (CNN) is the state-of-the-art technology used in computer vision. CNN models are compelling in computer vision tasks because the trainable convolutional filters can learn patterns in a visual field. They can be trained to recognize objects in an image. For training, the input is labeled images in RGB (red, green, blue) channels or in grayscale, and a trained model predicts the intended target class or location of the object. This study uses strain rate images to train models that can identify and locate fracture-hit events.

Many image classification applications utilize well-known pre-trained. In fact, we find a simple CNN architecture sufficient. The model consists of two convolutional layers followed by two fully connected flat layers and a sigmoid output for binary classification. Each convolutional layer includes convolution, batch normalization, ReLU activation, and max pooling.

Event Localization. In addition to identifying whether an image contains a fracture-hit event, we also want to know the location of the event. This is considered an object localization problem, practically a regression problem with the task of identifying where in the image the object of interest is in terms of time and depth. The location is typically characterized by a 2D coordinate, and width and height that define a bounding box around the object. In this application, we are only concerned with the 2D coordinate, hence the model has two outputs: location in X-axis (time) and in Y-axis (depth). The architecture of the localization model is the same the classification model trained to identify the presence of the event, except that the sigmoid output is replaced with two linear activations (essentially no activations). Correspondingly, we use mean squared error as the loss function since the training data are generated with well-defined distributions.

Results and Observation. The developed approach was trained and validated by datasets. The results are with satisfaction.

This study demonstrated the efficiency and accuracy of using convolutional neural network models to classify and localize fracture-hit patterns in simulated strain rate data. Both classification and localization models achieved near-perfect predictions. The results prove the feasibility of using CNN for real-time event detection from low-frequency DAS data. Because of CNN's flexibility and trainable nature, its applications extend beyond identifying strain rate patterns from low-frequency DAS data. This functionality can be developed as software packages for fiber optic sensing deployment, creating additional value for this already versatile technology. Edge detection is another plausible technique for locating fracture-hits, but it relies on assumptions about the image. The cone-shaped pattern must be smoothly and sharply defined. These assumptions often do not hold for field DAS data. The comparison of edge detection and CNN further supports the need for using CNN for image-based real-time event detection applications from fiber optic sensing data.

630515 Diagnosis of Multi-Stage Hydraulic Fracturing Using DAS and DTS Measurements

Abstract:

Distributed Temperature Sensing (DTS) and Distributed Acoustic Sensing (DAS) during hydraulic fracturing treatments are used to estimate fluid volume distribution among perforation clusters. DAS is sensitive to the acoustic signal induced by fluid flow through perforation holes during pumping of a stage, while DTS is sensitive to temperature variation caused by fluid flow inside a wellbore and in a reservoir.

Raw acoustic signal has to be transferred to frequency band energy (FBE) which is defined as the integration of the squared raw measurements in each DAS channel location for a fixed period of time. In order to be used in further interpretation, FBE has to be averaged between several fiber-optic channels for each cluster on each time step. Based on this input, DAS allows us to consider fluid flow through perforation stage by stage during an injection period, and to evaluate the volume of fluid pumped in each cluster location as a function of time, and therefor to estimate the cumulative volume of fluid injected into each cluster. This procedure is based on a lab-derived and computational dynamics model confirmed correlation between the acoustic signal and the flow rate. At each time step, we apply the perforation/fracture noise correlation to determine the flow rate into each cluster, constrained by the requirement that the sum of the flow rates into individual clusters must equal the total injection rate at that time.

On the other hand, the DTS interpretation method is based on the transient temperature behavior during the fracturing stimulation. After shut-in of a stage treatment, temperature recovers at a rate depending on the injected volume of fluid at the location. The interpretation procedure is based on the temperature behavior during the warm-back period. This temperature distribution is obtained by solution of a coupled 3-D reservoir thermal model with 1-D wellbore thermal model iteratively.

Once we confirm that the DAS and DTS interpretation methods provide comparable results of the fluid volume distribution, either of the interpretation results can be used as a known input parameter for interpreting the other measurand to estimate another unknown such as one of the fracture properties. In this work, the injected fluid volume distribution obtained by the DAS interpretation is used as an input parameter for a forward model which computes temperature profile in a reservoir. By conducting temperature inversion to reproduce the DTS data with the fixed injection rate for each cluster, we can predict how created hydraulic fractures are distributed along a wellbore as well as one of their properties. The inversion shows that multiple fractures are created in a swarm pattern from each perforation cluster with a much tighter spacing than the cluster spacing.

Using the field data provided by the Marcellus Shale Energy and Environmental Laboratory as an example, this study demonstrates that the DAS/DTS combined interpretation can be one of the valuable means to evaluate the fracturing treatment design and further understand the field observation of hydraulic fractures. The estimated fracture distribution is statistically characterized and compared with several completion parameters. This statistical analysis shows that the total fluid injection rate influences the fracture distribution in the example well.

625373 Refracturing in the Bakken – an Analysis of Data from across North Dakota

Abstract:

The Energy & Environmental Research Center (EERC), in close coordination with several of Williston Basin's premier operating companies, is engaged in a research program called the "Bakken Production and Optimization Program," or "BPOP," with the goal of simultaneously improving oil recovery in the North Dakota portion of the Bakken petroleum system (Bakken) while reducing its environmental footprint. BPOP is a consortium-funded program, with funding provided from a combination of the North Dakota Industrial Commission (NDIC) Oil and Gas Research Program (OGRP), several operating companies, and the U.S. Department of Energy.

One of the activities conducted by the EERC through BPOP is an evaluation of refracturing performance in the North Dakota portion of the Bakken. In this manuscript, refracturing refers to an operation to restimulate a well after an initial period of production, with the intent of either restoring or increasing well productivity and extending the productive life of a well. More than 90% of the estimated oil-in-place in unconventional reservoirs is not produced via primary production. Refracturing provides one technique for improving oil recovery by counteracting the effects of suboptimal completions of the original well such as undersized fracture treatments, suboptimal spacing of fracture intervals, inoperable completion failures, and damage mechanisms that cause a loss of connectivity to the reservoir.

Refracturing is a recent development for the Bakken but is not a new technique, as refractured wells, or refracs, were first implemented in the 1970s. One of the earliest refracturing operations in the Williston Basin was successfully performed in the Elm Coulee Field of Montana, where single-stage horizontal completions were restimulated. In North Dakota, Marathon Oil Company (Marathon) had an active refracturing program targeting older, openhole (barefoot) wells that were originally completed from 2006 to 2009 (classified as Generation 0 wells) in the Bakken. Marathon noted success with refracturing the Generation 0 wells; however, it also noted technical and operational challenges associated with later-generation wells having more than one completion stage. Whiting Oil and Gas Corporation reported the successful refracturing of a multistage horizontal well in the Bakken. The North Dakota Pipeline Authority (NDPA) identified and compiled all of the Bakken wells that were potentially refractured in North Dakota and evaluated production performance of those wells. The NDPA findings indicated that there were approximately 185 refracs across the Bakken (less than 2% of the total wells) and incremental oil production following refracturing was approximately 200,000 stock tank barrels (stb) per well. These estimates provide an example of the potential increased oil production associated with refracturing in the Bakken.

However, the oil and gas companies operating in the Bakken are cautious about broadly implementing refracturing because of the technical and operational challenges associated with the refracturing process. These challenges create uncertainty in postrefrac performance, which translates to elevated financial risk that limits access to investment capital, despite other economic incentives such as existing infrastructure (pipeline/facilities), a stable environment because the well already exists, known well details and intelligence (understanding of well operation best practices), and lower associated cost.

The goal of this BPOP study was to quantify uncertainty in refrac performance by investigating a large data set of refracs in the Bakken; 272 wells have been refractured in the North Dakota portion of the Bakken as of May 2019, which constitutes the data set used in the present study. The data set was compiled from various data sources including the North Dakota Geological Survey (NDGS), NDIC and FracFocus, the national hydraulic fracturing chemical registry. The data set includes the monthly oil production data as well as information specific to the original and refrac treatments for each of the 272 wells. In addition, decline-curve analysis was used to supplement these data and generate additional metrics for the economic analysis. The data set inventory revealed that the number of refracs increased considerably over the 2017–2018 timeframe, with 104 refracs conducted in 2017 and 2018, as compared with 168 refracs from 2009 through 2016, indicating the increased interest of oil companies in refracturing operations in the Bakken.

This manuscript discusses the results of 1) an evaluation of production performance of wells that have been refractured in the North Dakota portion of the Bakken; 2) an analysis of the impact of refracturing on offset wells (secondhand refracturing); 3) an assessment of refrac economics using a decision-tree modeling approach; and 4) an assessment of the potential economic impact of a broad refracturing program applied throughout the Bakken.

The production analysis of the refracs included several metrics to evaluate refracturing performance, including 1) a comparison of pre- and postrefracturing peak oil rates, 2) a change in daily production data following the refracturing operation, 3) a change in gas-oil ratio (GOR), and 4) incremental estimated ultimate recoveries (EURs) (i.e., the change in EUR between pre- and postrefracturing forecasts). In addition to summarizing the results across all refractured wells, the refracturing production performance was compared between wells that were originally completed as barefoot vs. nonbarefoot wells. Barefoot completions are believed to be understimulated during the original treatment and to have a greater undrained area and larger remaining hydrocarbon volume, resulting in commensurately better post-refracturing production performance. Furthermore, refracturing conducted on wells with barefoot completions are technologically easier and less costly to implement, which makes this group of wells potentially the best candidate for refracturing. The results showed that refracturing performed on barefoot wells outperformed those performed on nonbarefoot wells. For example, the median incremental EUR for the barefoot and nonbarefoot wells under the expected scenario was 340 and 175 thousand stb (Mstb), respectively. The postrefracturing uplift in daily oil production was 421 stb in barefoot completions as compared to 280 stb in nonbarefoot completions. The average GOR of the barefoot wells decreased by 10% on 90-day average daily production data, which provides supporting evidence that the refracturing provided access to previously undrained portions of the reservoir. Conversely, the average GOR of the nonbarefoot wells remained the same after refracturing.

Refracturing can not only restore or improve production of the well that is refractured but also improve production of offset wells located near the refractured well. An analysis of secondhand refracturing was performed to evaluate the production of wells that were offset from the refractured wells. The result showed that refracturing significantly impacted offset wells, and an average increase of 65 stb/day in daily oil production rate was observed in more than half of the offset wells analyzed after nearby refracturing.

An economic analysis was performed that focused on *discounted net oil revenue*, defined as the oil revenue after deducting royalties (16.67% applied pretax) and state tax (10% applied after deducting royalties), assuming an annual discount rate of 10% per year. A decision-tree modeling approach was used to evaluate the overall economic performance given a set of probabilities assigned to each potential outcome. Comparing the economic performance of wells that were originally completed as barefoot wells to nonbarefoot wells showed that barefoot wells had greater discounted net oil revenue. For example, the median discounted net oil revenue for barefoot wells was $4.9MM (million), while the median for nonbarefoot wells was $1.7MM, or approximately 65% lower.

An initial screening of existing wells in the Bakken identified 400 wells as potential candidates for refracturing using a simple set of criteria, specifically wells that had single-stage completions, older completion dates, and barefoot completions. The potential economic impact from refracturing all of these candidate wells resulted in an estimated discounted net oil revenue of approximately $2 billion, which reflects the median outcome after deducting the refracturing cost, taxes, and royalties. These results show that there is significant production potential of refracturing in the Bakken, particularly if single-stage, older barefoot wells are targeted.

625627 The Pore–Scale Darcy–Brinkman–Stokes Model for Iron–Precipitation in Shale Reservoir with the Injection of Hydraulic Fracturing Fluid

Abstract:

It has been reported that approximately 60% of total U.S. oil production comes from shale reservoirs. Hydraulic fracturing is widely used in shale reservoir productions to increase permeability. Hydraulic Fracturing Fluid (HFF) contains diverse chemical additives. When these additives mix with formation water, chemical reactions occur, resulting in the dissolution and precipitation of minerals. Previous experimental investigation indicated that the oxidation of released Fe^{2+} and subsequent precipitation of Fe^{3+} are the main paths of iron scale formation. pH is the most important factor affecting the release of Fe into solution. When the initial pH of solution was 2.0, the mineral precipitation occurred. However, when the reactors didn't contain the acid addition (initial pH 7.1), there was no detectable Fe in the solution. Most HFF additives extracted bitumen from oil and gas shales in the field, resulting in the liberation of bitumen, which formed a Fe^{3+}–bitumen complex and then decoupled bitumen and Fe (III) as free aqueous species. These subsurface chemical processes negatively impacted the hydrocarbon production by limiting transport. In this study, a new pore–scale Darcy–Brinkman–Stokes (DBS) model coupled with Level Set Method (LSM) is developed to address the microscopic phenomena during the iron–HFF interaction by numerically describing mass transport, chemical reactions, and pore structure evolution.

The new model has been developed based on OpenFOAM, which is an open source platform for computational fluid dynamics. Here, the DBS momentum equation was used to solve velocity equation by accounting for the fluid–solid mass transfer; an advection–diffusion equation was used to compute the distribution of injected HFF and iron. The reaction–induced pore evolution was captured by applying the LSM, where the solid–liquid interface was updated by solving level set distance function and reinitialized to a signed distance function. Then, a smoothened Heaviside function gave a smoothed solid–liquid interface over a narrow band with a fixed thickness. The stated equations were discretized by the finite volume method, while the reinitialized equation was discretized by the central difference method. Gauss linear upwind scheme was used to solve the level set distance function, and Pressure–Implicit with Splitting of Operators (PISO) method was used to solve the momentum equation. To validate the numerical method, we conducted the HFF reactive transport experiment with a shale sample. A sample from Marcellus shale was first soaked into HCl solution to remove carbonate and dried. Then, the sample was embedded in a single straight polydimethylsiloxane (PDMS) channel. The freshly made HFF, mainly containing organic additives, was injected into the system by a polytetrafluoroethylene (PTFE) tube from the left–hand side of the PDMS channel. A syringe pump provided the constant fluid velocity. PH value was adjusted by hydrochloric acid (HCl) and sodium hydroxide (NaOH) and set as 2.0 in the experiment. Injected velocity was 6×10^{-4} m/s; microchannel was 40 mm long, and the cross section of the channel was 1.5 mm × 0.3 mm. We weighted the sample before and after the reaction for four weeks. After the reaction, cross sections of the samples were characterized using Scanning electron microscope (SEM). The oxidation of Fe^{2+} was measured by ion chromatography with time intervals. The normalized sample mass was compared with our numerical results, showing good agreement between the two methods.

From the experiments, we found that total Fe and Fe^{2+} increased first and then decreased, suggesting the precipitation of Fe. The precipitation of Fe depends on pH value, carbonate contents, and extraction of bitumen. We found that under our experimental condition, the iron precipitation was not symmetrical. The iron tended to precipitate at the downstream side of the sample. Sensitivity analysis was conducted with various Damkohler number (Da_{II}) and Peclet number (Pe) between HFF and shale sample. We categorized the Fe (III) precipitation into three patterns as a function of Da_{II} and Pe: symmetrical smoothed growth, unsymmetrical growth, and dendritic growth. Pe and Da_{II} significantly affected the location of precipitation, which was critical in determining the injection parameters of hydraulic fracturing. When $Da_{II}<1$, the precipitation uniformly occurred on the solid surface both in upstream and downstream directions. When $Da_{II}>1$, the precipitation mainly occurred on the solid surface in upstream direction. When $Pe>1$, Fe (II) was transported deeply into and precipitated inside the pores. When $Pe<1$, the precipitation of Fe (III) occurred mainly on the solid surface in upstream direction, and they were easily precipitated inside the small pore structures. The porosity–permeability relationship was subsequently presented. We also simulated this process under different mineralogical samples from digital rock images and obtained the following interesting findings. When the samples contained the large amount of calcite, no precipitation was observed. However, when the content of calcite was low and the content of dolomite was high, we could observe slight precipitation.

The newly developed pore–scale model allows high confidence in the description of Fe (II) dissolution, transport, and Fe (III) precipitation, which has improved our understanding of mineral dissolution and precipitation in shale during hydraulic fracturing. The model shows fast convergence while requiring low computational load. The results can provide reliable guidance for injecting HFF in shale reservoirs to avoid clogging and wellbore pollution. Understanding the precipitation of Fe (III) and the release and transport of Fe (II) will lead to a highly efficient hydraulic fracture project.

APPROXIMATION ANALYSIS OF PITZER ACENTRIC FACTOR BY MOLECULAR DYNAMICS

***Bin Ju*[1*], *Ping Guo*[1], *Yizhong Zhang*[2], *Long Yang*[2]**
*Corresponding author. Email address: bc_j2009@foxmail.com
[1] School of Oil and Natural Gas Engineering, Southwest Petroleum University, Chengdu, Sichuan, China, 610500
[2]Hubei Cooperative Innovation Center of Unconventional Oil and Gas, Yangtze University, Hubei, Wuhan, China, 430100

Abstract

In daily phase behavior analysis, the Pitzer acentric factor has been used to measure the difference between equation of state and the real physical result. However, considering that this factor was defined by experimental results, it may be invalid on the surface of an extremely irregular boundary microcosmic conditions. In the research where accurate result on irregular boundary needs to be described, the accurate reorganization process must be taken into consideration. This study uses the statistical method to build a mathematical model which describes the behavior of a definite molecule and uses stochastic process and sensitivity analysis to describe how the surroundings affect this molecule. The solution of this mathematical model proves that the Pitzer acentric factor is approximated by a linear function of logarithmic pressure. Besides, this model extends a constant Pitzer acentric factor to an area of approximate inner convergence.

Keywords: Pitzer acentric factor, irregular boundary, molecule behavior, game theory, mixed Nash equilibrium

1. Introduction

The Pitzer acentric factor is defined by Pitzer [1] which has been used to measure the difference between equation of state and the real physical result. When molecule has a geometry construction significant different from perfect ball there will be error for the solution of equation of state. Pitzer solved this problem by experimental and experienced method. However, it can't be used to describe some extremely unregular boundary flow behavior, which is certificated by a group of research [2-6] by different methods that there is a significate difference from phase behavior in macrocosmic flow.

Methods of this research is using statistical method to build a mathematical model which describes the behavior of a definite molecule, and then using stochastic process and sensitivity analysis to describe how the surroundings affect this molecule. When we solved these models partly, we used hypothesis test to clarify that the result has no significant difference from the experimental result.

2. Assumptions

Fluid flow and its asymptotic polynomial functions are almost surely uniformly continuous. This assumption is equivalence to that there is unsignificant influence whenever delete any Zero-measured part from system.

Phase behavior is only decided by 3-dimensional Euclidean distance of molecule and the average time for molecules elastic collision with each other. This assumption is used to define a boundary between different phase, so it can help us to define an almost surely tendency that some specific point located in the decision space.

We don't consider 3 following elements: the area of critical point where may not have a significant boundary, chemical reaction where may change molecule itself and non-equilibrium behavior because of microcosmic behavior is considered by a different direction.

3. Methods

3.1. One-dimensional Flow

Under these assumptions, the mathematics works start with the simplest model that the macrocosmic and microcosmic are both 1-dimensional. Consider 2 special direction of microcosmic movements that microcosmic has the largest and the smallest inner product with the macrocosmic flow.

Considering the simplest model: Brownian movement of molecule only on one-dimensional flow. Using Functions of Complex Variables to describe this behavior:

For a continuous flow, the equation of flow is

$$F(x, v) = 0$$

Extend the flow behavior to complex plane and take microcosmic movement into consideration, the equation of flow is

$$F(x + c\varepsilon, v) = 0$$

For any complex plane, unit real vector and unit imaginary vector are a group of orthonormal bases. In another words

$$\forall \varepsilon, \exists x_r, x_i \;\; s.t. \;\; \varepsilon = x_r + ix_i$$

Besides, any real vector is orthogonal with any imaginary vector, the equation of flow can be written as

$$F(x+cx_r, v+v_r)+G(x_i, v_i)=0$$

Any continuous function fits intermediate value theorem, which means

$$\exists x_s \in \dot{B}(x), s.t.\ \ F(x,v)+G(cx_r+x_i, v_r+v_i)=0$$

Or

$$\exists x_s \in \dot{B}(x), s.t.\ \ F(x,v)=G(cx_r+x_i, v)$$

Despite sign because $G(\cdot)$ is arbitrary. What's more, delete any zero-measure set from a certain probability space did no significant influence on the convergence in probability.

For the right-hand side of the equation of microcosmic flow is nothing related with x, it is convergence in probability with some nonhomogeneous equation of macrocosmic flow.

Back to the assumption, we have assumed that phase behavior is only decided by 3-dimensional Euclidean distance of molecule and the average time for molecules elastic collision with each other.

In other words, define L as the distance between one molecule and the nearest other molecule, for any molecule movement speed $v=v_0$, there is a maximum distance of vapor $L_{vmax}=L_{lmin}$ which is the minimum distance of liquid. Denote this critical distance as $L_c(v)$, then the event representation function

$$I\{behavior\ is\ changed\}=\begin{cases}1 & case\ when\ (L(v_0)-L_c(v))(L(v_1)-L_c(v))<0\\ 0 & case\ when\ (L(v_0)-L_c(v))(L(v_1)-L_c(v))>0\end{cases}$$

Besides, when

$$L(v_0)-L_c(v)\ =0$$

or

$$L(v_1)-L_c(v)=0$$

It means that corresponding located in the hybrid of two-phase system.

Then the question changes to solve $L_c(v)$, considering another assumption: The area of critical point and its neighborhood is not considered.

It is the same with making a hyperplane to divide the space apart to two parts. There is significant difference of behaviors, so there must be significant difference between models, in another word, these two behaviors can't fit any model at the same time.

There are two samples can't pass the Chow Test. Using these rules there must be Only one $L_c(v)$.

3.2. Two-dimensional Flow under Cylindrical Coordinates System

However, the 1-dimensional flow can't be approximate enough so we must extend this model to higher dimension. Considering the simplest 2-dimensional flow combination: the macrocosmic flow is 1-dimensional, and the microcosmic plane is orthogonal with the macrocosmic flow. The flow control system is a cylinder in 3-dimensional space.

Firstly, we choose cylindrical coordinates to describe the 2-dimensional flow. It can be calculated that for the extremely close-packed array, circle with a same radius can cover the space as the maximum ratio of

$$\frac{\pi}{4\sqrt{3}}\approx 0.45345=45.345\%$$

For a uniformly system, the radius must be the same. So, static models can't fit well in polar coordinates. It must be solved by dynamic models.

For a continuous flow, the equation of flow is

$$F(x, v) = 0$$

Extend the flow behavior by taking microcosmic movement into consideration, the equation of flow is

$$\begin{cases} F(x, v) = 0 \\ G(x; r, \theta; v_r, v_\theta) = 0 \end{cases}$$

It is the same with 1-dimensional flow, linear combination of microcosmic and macrocosmic can't cover the entire space. In other words

$$\begin{cases} F(x, v) = 0 \\ G(x; r, \theta; v_r, v_\theta) = 0 \end{cases} \Leftrightarrow \{(x, r, \theta, \vec{v}) | (x, r, \theta, \vec{v}) \in S\}$$

Needs more smaller radius models to cover.

For previous discussion, circle can't cover the space at the same radius even in extremely close-packed array.

If we want to use different radius, it is easy to find out by using geometry knowledge that the regular array has the most efficiency as the partial summation of infinite series

$$E = \sum_{i=1}^{N} (1 - ra)^n$$

Where ra is the attenuation coefficient means every inferior cylinder will reduce the ratio of ra. Obviously

$$E \to 1 \; when \, n \to \infty$$

Or

$$\exists M \in N, \forall \varepsilon > 0, \exists \delta = f(\varepsilon) \; s.t.$$

$$\int \| \{(x, r, \theta, \vec{v}) | F(x, v) = 0, G(x; r, \theta; v_r, v_\theta) = 0\} - \{(x, r, \theta, \vec{v}) | (x, r, \theta, \vec{v}) \in S\} \| \, d\theta < f(\varepsilon)$$

Denote

$$D = \| \{(x, r, \theta, \vec{v}) | F(x, v) = 0, G(x; r, \theta; v_r, v_\theta) = 0\} - \{(x, r, \theta, \vec{v}) | (x, r, \theta, \vec{v}) \in S\} \|$$

Building a simpler model, fix an origin located in the center of unit ball, denote as O. Discuss the situation that collecting n points on the boundary of cylinder, while m points have been collected.

Denote these m fix points as $X_1, X_2, \cdots, X_m$, then $\overrightarrow{OX_1}, \overrightarrow{OX_2}, \cdots, \overrightarrow{OX_m}$ define a m-dimensional subspace W.

Considering X_{m+1} which is the point $m + 1$, divide position vector of this point into vertical component and horizontal component as

$$\overrightarrow{OX}_{m+1} = \boldsymbol{r}_\perp + \boldsymbol{r}_W; \boldsymbol{r}_W \in W, \boldsymbol{r}_\perp \in W^\perp$$

To solve $|\boldsymbol{r}_\perp|$.Considering X_{m+1} located inside the ball rather than uniformly distributed on the sphere. For any

$$|\boldsymbol{r}_\perp| = r < 1, |\boldsymbol{r}_W| < \sqrt{1 - r^2}$$

The volume element $d\boldsymbol{r} = d\boldsymbol{r}_\perp d\boldsymbol{r}_W$. Define population as uniformly distribution in $B_{n-m} \otimes B_m$, then

$$P(r < |\boldsymbol{r}_\perp| < r + dr, |\boldsymbol{r}| < 1) = P_{n-m}(r < |\boldsymbol{r}_\perp| < r + dr) P_m\left(|\boldsymbol{r}_W| < \sqrt{1 - r^2}\right)$$

Or

$$P(r < |\mathbf{r}_\perp| < r + dr, |\mathbf{r}| < 1) \propto f_{n-m}(r) dr F_m\left(\sqrt{1 - r^2}\right)$$

Besides,

$$P(r < |\mathbf{r}_\perp| < r + dr, |\mathbf{r}| < 1) = n r^{n-m-1} (1 - r^2)^{m/2} dr$$

So, the probability function of r is

$$f_{n,m}(r) \propto r^{n-m-1} (1 - r^2)^{m/2}$$

The probability function of r^2 is

$$f_{n,m}(r^2) \propto (r^2)^{\frac{n}{2} - \frac{m}{2} - 1} (1 - r^2)^{\frac{m}{2}}$$

Using the integral of any probability function identically equal to 1, r^2 following and only can be following Beta distribution:

$$\mathrm{r}^2 \sim \mathrm{Beta}\left(\frac{\mathrm{n} - \mathrm{m}}{2}, \frac{\mathrm{m}}{2} + 1\right)$$

$$\overline{|\mathbf{r}_\perp|} = E_{r^2 \sim \mathrm{Beta}\left(\frac{n-m}{2}, \frac{m}{2}+1\right)}\left[(r^2)^{1/2}\right] = \frac{\Gamma\left(\frac{n}{2} + 1\right)\Gamma\left(\frac{n - m + 1}{2}\right)}{\Gamma\left(\frac{n + 3}{2}\right)\Gamma\left(\frac{n - m}{2}\right)}$$

Transfer this result into the situation that X_{m+1} located on the sphere rather than inside the ball.

For function $g(x)$, if

$$g(\boldsymbol{x}) = g_r(|\boldsymbol{x}|) g(\widehat{\boldsymbol{x}}), \forall \boldsymbol{x} > 0$$

Which means that radial direction of this function is independent with angle direction, then

$$E_{\boldsymbol{x} \sim U(B_n)}[g(\boldsymbol{x})] = E_{\boldsymbol{x} \sim U(B_n)}[g_r(|\boldsymbol{x}|)] E_{\boldsymbol{x} \sim U(S_n)}[g(\boldsymbol{x})]$$

For $g(\boldsymbol{x}) = |\boldsymbol{x}_\perp|$, always exists $g_r(\boldsymbol{x}) = \boldsymbol{x}$

$$E_{\boldsymbol{x} \sim U(B_n)}[g_r(|\boldsymbol{x}|)] = \int_0^1 x f_n(x) dx = \frac{n}{n + 1}$$

$$\overline{|\boldsymbol{r}_\perp|} = E_{\boldsymbol{x} \sim U(S_n)}[g(\boldsymbol{x})] = \frac{\Gamma\left(\frac{n}{2}\right)\Gamma\left(\frac{n - m + 1}{2}\right)}{\Gamma\left(\frac{n + 1}{2}\right)\Gamma\left(\frac{n - m}{2}\right)}$$

Because of m is arbitrary, take the linear recurrence relation and the conclusion is that when molecule is on the boundary of phase behavior, where has a significant difference between molecule microcosmic behavior, denote $F(\cdot)$ as probability distribution function of this molecule located where the molecule located in vapor is equivalent to the molecule located in liquid, and $F(\cdot)$ satifies

$$F^{-1}(x_1) F^{-1}(x_2) \propto F^{-1}(x_1 + x_2)$$

Considering that

$$D = \|\{(x, r, \theta, \vec{v}) | F(x, v) = 0, G(x; r, \theta; v_r, v_\theta) = 0\} - \{(x, r, \theta, \vec{v}) | (x, r, \theta, \vec{v}) \in S\}\| \to 0$$

Or the form of integration

$$\int Dd\theta = \int \|\{(x,r,\theta,\vec{v})|F(x,v)=0, G(x;r,\theta;v_r,v_\theta)=0\} - \{(x,r,\theta,\vec{v})|(x,r,\theta,\vec{v}) \in S\}, \theta\| d\theta \to 0$$

3.3. Two-dimensional Flow under Cartesian Coordinates System

For a continuous flow, the equation of flow is

$$F(x,v)=0$$

Extend the flow behavior by taking microcosmic movement into consideration, the equation of flow is

$$\begin{cases} F(x,v)=0 \\ G(x;y,z;v_y,v_z)=0 \end{cases}$$

It is the same with 1-dimensional flow, linear combination of microcosmic and macrocosmic can cover the entire space. In other words

$$\begin{cases} F(x,v)=0 \\ G(x;y,z;v_y,v_z)=0 \end{cases} \Leftrightarrow \{(x,y,z,\vec{v})|(x,y,z,\vec{v}) \in S\}$$

4. Results and Discussions

In sum

$$\begin{cases} F(x,v)=0 \\ G(x;r,\theta;v_r,v_\theta)=0 \end{cases} \Leftrightarrow \begin{matrix} F(x,v)=0 \\ G(x;y,z;v_y,v_z)=0 \end{matrix}$$

Which means factor or boundary where there is nothing difference between strategies fits

$$F^{-1}(x_1)F^{-1}(x_2) \propto F^{-1}(x_1+x_2)$$

Or

$$f(x) = \frac{f(x_1)f(x_2)}{f(x_1+x_2)}\ln(x)$$

Where x_1, x_2 are any two points in domain of definition, $f(\cdot)=F^{-1}(\cdot)$.

The real flow behavior can be combination of the previous 1-dimensional and 2-dimensional flow.

For the same analysis logic, $L_c(v)$ defined by Chow Test. While in 3-dimensional flow because the microcosmic and macrocosmic movements are all 3-dimensional, the result needs divide the combinatorial number. Besides, for any macrocosmic, there must only one direction. Using this direction be basic axis and there are 2 axes left for microcosmic movements.

So, the limitation function of pure substance of Pitzer acentric factor can be convergence with probability by

$$\mathcal{L}(s) = \int_{r_1}^{r_2} r^s lnr dr \Big/ \int_{r_1}^{r_2} r^{s+1} lnr dr$$

Where r_1, r_2 satisfies

$$minimum \int_0^{r_1} lnr dr, s.t. \int_0^{r_2} lnr dr = 0$$

The Pitzer acentric factor is exponential function which radix is the limitation function $\mathcal{L}(s)$ and exponent varied for function of s, $\mathcal{E}(s)$. Under the assumptive condition

$$s \in [2,3], \mathcal{E}(s) = s^{-1}$$

The Pitzer acentric factor coefficient located in

$$p_e \in [0.6433, 0.7452]$$

The Pitzer acentric factor is

$$\mathcal{P} = \lg(T_r | p_r = p_e) - 1$$

5. Conclusion

The Pitzer acentric factor calculated in this article is difference from the experience result, for the experience result gives 0.7 constant while this article shows that it should be a random value of 0.6433 and 0.7452, it depends on how the boundary layer cover the whole flow.

This article takes no consideration that laws of linear combination. More information needs from experiments if we want to calculate Pitzer acentric factor more accurate, which aiming at how to matching the weight for different condition including temperature, pressure, and boundary of flow.

Even it is not accurate enough for factor calculation, this article provides a method how to qualitative solution sensitivity of Pitzer acentric factor, which can provide a reference basis for description of irregular boundary phase behavior.

Acknowledgements

The authors would like to acknowledge the support from National Natural Science Foundation of China (NSFC Grant No.52004032).

References

[1] Pitzer, K. S.. (1955). The volumetric and thermodynamic properties of fluids. I. theoretical basis and virial coefficients. Journal of the American Chemical Society, 77(13), 3427-3433.

[2] Tu, H. , Guo, P. , Jia, N. , & Wang, Z. . (2018). Numerical evaluation of phase behavior properties for gas condensate under non-equilibrium conditions. Fuel, 226, 675-685.

[3]Eilerts, Kenneth C . Integration of Partial Differential Equation for Transient Radial Flow of Gas-Condensate Fluids in Porous Structures[J]. Society of Petroleum Engineers Journal, 1965, 4(04):291-306.

[4] Muskat M M . Physical Principles of Oil Production[M]. McGraw-Hill Book Co. 1949.

[5] Al-Wahaibi Y M , Muggeridge A H , Grattoni C A . Gas–oil non-equilibrium in multi-contact miscible displacements within homogeneous porous media[J]. Journal of Petroleum Science & Engineering, 2009, 68(1-2):71-80.

[6] Al-Wahaibi Y , Muggeridge A , Grattoni C . Gas/Oil Nonequilibrium in Multi-contact Miscible Displacements Within Homogeneous Porous Media[C]// SPE/DOE Symposium on Improved Oil Recovery. 2006.

628425 Unravelling Catalysis By Small Metal Carbide Clusters for Methane Dehydroaromatization

Abstract:

Direct conversion of methane to aromatics in non-oxidative conditions generally utilizes a bifunctional catalyst having active molybdenum species supported on zeolite (Mo/HZSM-5). In this reaction, the nature of the active site for methane activation on Mo is suggested to be a carbide (or an oxycarbide) and the catalyst is known to deactivate from carbon deposition over time due to the formation of polyaromatic hydrocarbons. Herein, density functional theory (DFT) calculations are utilized to study the catalytically active carbide (Mo_xCy) clusters in C-H bond activation and C-C coupling reaction to form C_2 intermediates[1]. A massively parallel cascade genetic algorithm (cGA) is applied to scan the potential energy surface (PES) for all possible low-energy structures (including the global minimum) of Mo_xCy. Catalytic activity of the global minimum and metastable clusters are accessed and compared for methane activation, highlighting the role of metastable clusters influencing the overall reaction rates[2]. A relatively new paradigm of accessing the reaction rates is promulgated which is incorporating an ensemble average representation of all active species of Mo_xCy.

Theoretical calculations suggest a key factor in deciding experimental protocols to control catalyst reactivity and stability. This involves monitoring the evolution of the molybdenum carbide species which dictates the anchoring of the carbide cluster on the zeolite support. The mechanistic insights obtained from DFT calculations are thus providing a molecular level engineering approach, wherein the catalyst is rationally synthesized to obtain desired active sites anchored on the zeolite support, which are stable in providing high aromatic selectivity and overall reactivity. This is implemented in experiments at three stages; a) by altering the Mo precursors used in the synthesis of Mo/HZSM-5 catalyst, b) treating the zeolite support with the boric acid to alter support acidity[3] and on c) changing the carburizing condition from pure methane to a mixture of hydrocarbon gases (mimicking the natural gas). In all three cases, the evolution and control of the active molybdenum carbide species, anchored on the zeolite support, is highlighted as a molecular level toolbox to achieve desired reactivity with a potential for higher catalyst stability.

References

1 T. S. Khan, S. Balyan, S. Mishra, K. K. Pant and M. A. Haider, *J. Phys. Chem. C*, 2018, **122**, 11754–11764.

2 S. Balyan, S. Saini, T. S. Khan, K. K. Pant, P. Gupta, S. Bhattacharya and M. A. Haider, *Nanoscale*, 2020

3 S. Balyan, M. A. Haider, T. S. Khan and K. K. Pant, *Catal. Sci. Technol.*, 2020, **10**, 3857–3867.

625543 A Study on Supported Bimetallic Catalysts Synthesis & Characterization for Direct Conversion of CO_2 and Methane into Value Added Chemicals

Abstract:

The growing demand for conversion of natural gas into value-added chemicals has been the focus of a significant number of studies[1, 2]. Specifically, the direct conversion of methane and CO_2 to higher-value oxygenates has substantial industrial value. Due to the thermodynamic stability of CO_2 and methane molecules, direct conversion to high-value product is commercially limited. The following is a two-step process for activation of methane and CO_2: the carboxylation of CH_4 to acetic acid (1), which is then reacted with acetylene to form vinyl acetate (2)

(1) $CO_2 + CH_4 \rightarrow CH_3COOH$

(2) $CH_3COOH + C_2H_2 \rightarrow$ vinyl acetate

This formation of acetic acid (1) is thermodynamically limited, but it can be addressed by coupling this reaction with a thermodynamically favorable reaction, e.g., reacting acetic acid with acetylene (2) to form vinyl acetate monomer (VAM), a high-value industrial intermediate. The overall reaction with acetylene ($CH_4 + CO_2 + C_2H_2 \rightarrow$ VAM) is thermodynamically more feasible than (1) ($\Delta G_r = +6.4$ kJ/mol). Previously, Spivey et al. demonstrated $CH_4 + CO_2 + C_2H_2 \rightarrow$ VAM between 200 and 400 ^{0}C at 1 atm [3], but VAM formation was not stable at these conditions.

This study focuses on design, synthesis and characterization of bimetallic catalysts for the conversion of methane and CO_2 into high value chemicals in a one-pot reaction. It has been reported[4] that metals like Au, Ag, Pd, Rh can activate methane and help couple it with CO_2 and acetylene. Therefore, TiO_2 supported bimetallic catalysts such as Rh-Au, Pb-Au, Ag-Au have been synthesized in-house followed by characterization via STEM-EDX, L edge XANES, XRD, and XPS analysis.

The bimetallic catalysts were prepared using deposition-precipitation method. 1 wt % of Rh, Ag, Pb, each were added to 1 wt% of Au and the metals were supported on TiO_2 for greater dispersion. After that, the catalysts were characterized to confirm the dispersion of metals and the proximity required to form the bimetallic sites. STEM confirmed that the metals were stationed adjacent to each other to create the bimetallic formation, while EDX confirmed the loading of the active metals on TiO_2. L edge XANES was performed on Au active sites to confirm the transformation of oxidation states as Au react with other metals (Pd, Ag) to form bimetallic sites, similar to the energy shift at white lines (encircled) observed in the literature for Au bimetallics (Figure 1). Successful formation of bimetallic sites was further evaluated via XPS analysis. XRD showed no crystalline structures in the fresh catalysts, which reveals the high dispersion of active metals on TiO_2 surface. The catalysts were later tested for direct activation of methane and CO_2 into oxygenates, which showed successful conversion to C_3 and C_4 products.

Reference:

[1] M.A. Abedin, S. Kanitkar, S. Bhattar, J.J. Spivey, Sulfated hafnia as a support for Mo oxide: A novel catalyst for methane dehydroaromatization, Catalysis Today, 343 (2020) 8-17.

[2] M.A. Abedin, S. Kanitkar, S. Bhattar, J.J. Spivey, Promotional Effect of Cr in Sulfated Zirconia-Based Mo Catalyst for Methane Dehydroaromatization, Energy Technology, 0 (2019) 1900555.

[3] J.J. Spivey, E.M. Wilcox, G.W. Roberts, Direct utilization of carbon dioxide in chemical synthesis: Vinyl acetate via methane carboxylation, Catalysis Communications, 9 (2008) 685-689.

[4] C. Palmer, D.C. Upham, S. Smart, M.J. Gordon, H. Metiu, E.W. McFarland, Dry reforming of methane catalysed by molten metal alloys, Nature Catalysis, 3 (2020) 83-89.

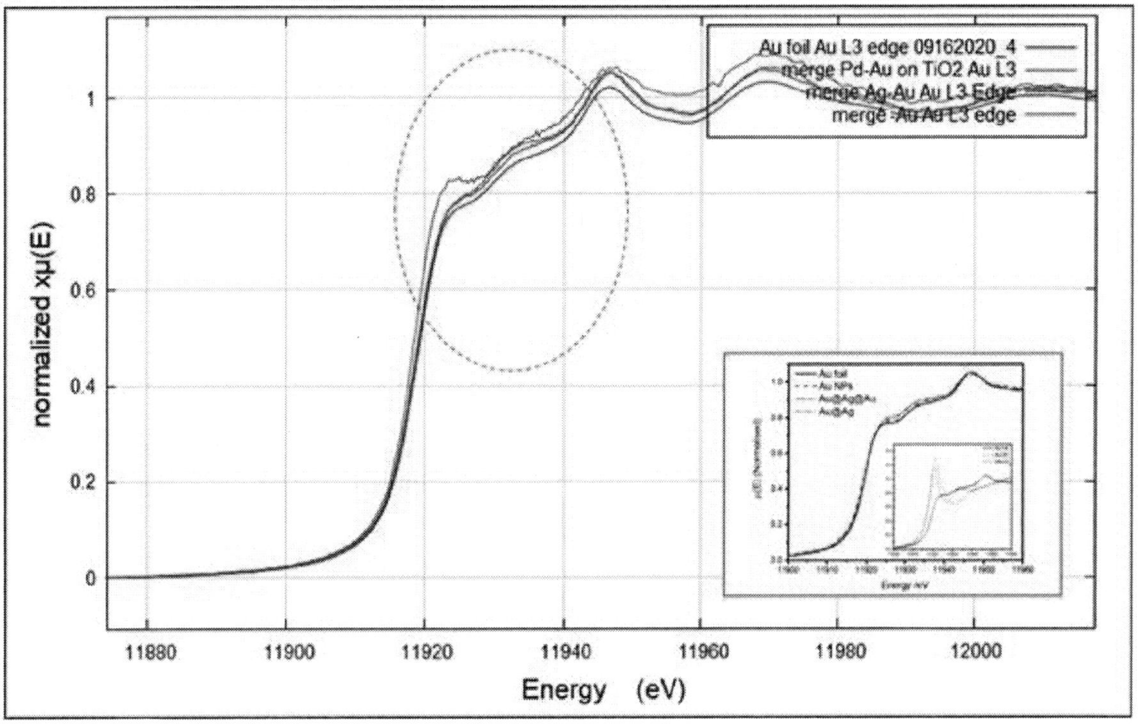

Figure 1: L-edge XANES studies on the prepared bimetallic catalysts: Ag-Au, Pd-Au, Rh-Au supported on TiO_2.

633401 Effect of Catalyst Pretreatment on Methane Conversion Under Microwave-Assisted and Conventional Thermal Method

Abstract:

The utilization of remote associated flare gas through conversion into value added products at well-sites presents an opportunity to provide value to an otherwise wasted resource and to remediate serious environmental and economic challenges facing oil and gas producers. A microwave-based process offers rapid, volume-based heating which can provide on-demand chemical production and easily load follow the heat demands of the dynamic flows coming out of the well. The non-oxidative conversion of methane is a process intensified chemical process to produce value-added chemicals and gases such as aromatics and hydrogen. However, challenges of catalyst rapid coking and thermodynamic limitations are always in its way towards commercialization. A better understanding for catalyst carburization is still needed to control its fast deactivation. In this work, Mo/HZSM-5 catalyst was used as the most studied catalyst in literature for this reaction. The catalyst was activated under different gases: CO, syngas, and methane using microwave-assisted and conventional thermal methods. The correlations between the product yields and selectivities were correlated with activation gas used and heating mechanism applied to carburize Mo species. Modeling and characterization work are presented to determine the activity-structure relationship. Multiphysics numerical modeling work focused on understanding the effect of reactor geometry on fluid flow and heat transfer and its possible relation to catalyst activation.

624407 Valorisation of Mining Residues for Direct CO2 Hydrogenation

Abstract:

Direct CO_2 hydrogenation combines in a single step the reduction of CO_2 to CO and the reverse water-gas-shift reaction (RWGS). This process arouses great interest for its role in the reduction of CO_2 emissions, the production of synthetic fuels and space exploration.

The use of synthetic liquid hydrocarbons requires no significant changes to the existing infrastructure and engines, because they can be processed to fit the current specifications. Direct hydrogenation of CO_2 (the main greenhouse gas) for the production of synthetic fuels does not require a gasification step, in addition to being more economical and environmentally friendly.

In this work several catalytic formulations, prepared from mining residues and doped active components such as ceria, zirconia and potassium have been chosen to be tested. These multifunctional catalysts provided more than 1 types of active sites, which play a synergetic role in the catalytic conversion of CO_2. The selectivity of the process in long carbon chains such as petrol, diesel and waxes, similar to those produced from fossil fuels, has been evaluated as function of the test conditions. Mechanistic understanding of the catalyst/support interactions gained from XRD and SEM analyses guided the tuning of the surface properties and provide insight towards maximizing yields.

624052 Statistical Reaction Analysis of Residue Desulfurization Reactor Based on Component Structures

Abstract:

Heavy oils contain many kinds of organic compounds which have complex structures. Thus, it is difficult for us to figure these characteristics to develop suitable catalysts. Recently, due to the development of Fourier transform ion cyclotron resonance mass spectrometry (FT-ICR-MS), the detailed composition data of heavy oils can be obtained.

In this research, the authors proposed a new analyzing method of chemical reactions using the input/output composition data from FT-ICR-MS. In this method, representative chemical reactions were defined and assumed the series of reactions linearly. As a result, the conversion ratios of each chemical reaction can be simply obtained and statistically compared by component structures.

In the next, the authors applied this method to a residue desulfurization reactor in petroleum refinery as a case study and compared the reaction ratios by component structures. From this analysis, the authors found that the component structure significantly affects to the conversion ratio in desulfurization reaction and capture the characteristics of this reaction. Therefore, this analysis might be a promising to analyze reaction for providing insights of catalysts and process design.

628723 Electrocatalyst Development for Electrochemical Oxidative Dehydrogenation of Ethane

Abstract:

With the development of unconventional resources, there will be an anticipated increase in natural gas production [1]. An increase in natural gas production incentivizes the development of methods to handle natural gas liquids, the hydrocarbon compounds in natural gas that are heavier than methane, such as ethane [2]. Contemporary natural gas liquid processing methods focus on producing or transporting the methane component of natural gas at significant economic and energy investment [3]. Work at Ohio University's Institute for Sustainable Energy and the Environment, supported by the U.S. Department of Energy's National Energy Technology Laboratory [DE-FE0031709], focuses on upgrading ethane by means of an electrocatalytic solid oxide fuel cell system that performs electrochemical oxidative dehydrogenation (e-ODH) to ethylene [4,5]. Under this project the team has synthesized and evaluated perovskite electrocatalysts to conduct e-ODH. Promising electrocatalyst compositions identified through this effort can be incorporated into the proposed modular processing technology that would enable economical ethane separation and upgrading.

Perovskite electrocatalysts were synthesized within a porous yttria-stabilized zirconia (YSZ) scaffold via infiltration, serving as the e-ODH anode of a solid electrolyte fuel cell using lanthanum strontium manganite (LSM20) as an oxygen reducing cathode electrocatalyst. The infiltrated anode electrocatalysts were characterized with scanning electron microscopy and X-ray diffraction. The ethane oxidation is catalyzed on the porous scaffold anode while the oxygen reduction occurs on the cathode, following the reaction chemistry shown below.

Anode: $C_2H_6 + O^{2-} \rightarrow C_2H_4 + H_2O + 2e^-$

Cathode: $1/2O_2 + 2e^- \rightarrow O^{2-}$

Overall ODH: $C_2H_6 + 1/2O_2 \rightarrow C_2H_4 + H_2O$

The infiltrated electrocatalysts' e-ODH capabilities were evaluated through electrochemical testing of the solid oxide fuel cell and quantitative gas chromatography analysis of the product gases under galvanostatic conditions.

This work has demonstrated the suitability of the electrocatalyst synthesis techniques and the favorability for ethylene selectivity over certain electrocatalyst compositions that will be presented in this talk. More broadly, this project has elucidated the influence of catalyst material upon ethane e-ODH chemistry and showed potential feasibility for incorporation into the modular design. In this talk electrocatalyst material characterization and fuel cell test results will be presented.

1. MacIntyre, S. U.S. Energy Information Administration: U.S. ethane production, consumption, and exports expected to increase through 2018, posted January 17, 2017. https://www.eia.gov/todayinenergy/detail.php?id=29572 (accessed October 31, 2020).
2. U.S. Energy Information Administration: What are natural gas liquids and how are they used?, posted April 20, 2012. https://www.eia.gov/todayinenergy/detail.php?id=5930 (accessed April 9, 2021).

3. Miller, T.; Beck, D. Gas Processing & LNG: Maximize propane recovery and ethane rejection at cryogenic gas plants. http://gasprocessingnews.com/features/201612/maximize-propane-recovery-and-ethane-rejection-at-cryogenic-gas-plants.aspx (accessed October 31, 2020).
4. Daramola, D.A.; Velraj, S.; Trembly, J. Electrochemical Oxidative Dehydration (e-ODH) As a Process Intensification Platform in Shale Gas Upgrading. In *2019 AIChE Annual Meeting*; Orlando, FL, November 10-15, 2019.
5. Kasick, A.; Hajer, A.A.; Daramola, D.A.; Velraj, S.; Trembly, J. Advances in the Use of Electrochemical Oxidative Dehydrogenation to Convert Natural Gas Liquids to Value Added Chemicals. In *2020 AIChE Annual Meeting*; San Francisco, CA, November 16-20, 2020.

622167 Reduction of NOx Emissions Via CFD-Based Optimization of a Steam Cracking Firebox

Abstract:

Introduction

Light olefins such as ethylene, propylene and butadiene are the key building blocks for all kinds of plastics and high-value chemicals, the production of which is driven by population growth and prosperity. Today, steam cracking of hydrocarbon feedstocks is the leading technology to produce these building blocks. As very high temperatures of up to 1130 K are needed for the reactions to occur, steam cracking takes place in tubular reactors suspended within large-scale fireboxes, i.e., furnaces fired by numerous floor and sidewall burners.

In industrial practice, it is often assumed that the temperature in a steam cracking firebox is completely uniform or at best changing according to the height of the furnace, hence disregarding the complex 3D heat transfer phenomena that take place. Adjustments to the firebox conditions are generally avoided during operation because of the difficulty to measure their impact. Only secondary measures like stack oxygen and NOx concentrations, bridgewall temperatures and coil outlet temperatures are monitored. Because steam cracking fireboxes are in fact far from uniform, it is impossible to pinpoint problem areas using only such discrete measurements that are typically located at the boundaries of the firebox. Bad burner management can cause excessive CO, NOx or soot formation due to suboptimal combustion conditions[1]. Additionally, flame rollover and impingement can occur[2], resulting in a non-uniform heat flux to the reactor coils and even hot spots[3]. The main possible handles to alter the performance of steam cracking furnaces are the air flowrates to the burners. Typically each burner has one or more air ducts to be able to adjust each burner separately or even adjust different air flowrates to a single burner. As more than 100 burners can be present in an industrial steam cracking furnace, this results in a huge amount of degrees of freedom during the operation of the furnace. Because of the high temperatures obtained when combusting fuel gas with air, thermal NOx formation is another significant issue in steam cracking furnaces. Furthermore, there is only a narrow window of combustion temperatures in which nearly full combustion is achieved while NOx formation remains low. This balance needs to be achieved in industrial burners, especially in high-temperature applications such as steam cracking furnaces.

Given the importance of the above considerations, in this contribution, 3D CFD simulations are used to evaluate the effect of altering the air distribution to the furnace. In particular the ratio of primary to secondary air towards the floor burners and the sidewall burners is investigated to evaluate what the optimum is to minimize NOx formation and maximize run length.

Methodology

The 3D firebox simulations are performed using the commercial CFD software Ansys FLUENT 19.0. Conservation equations for mass, momentum, species and energy are solved for a compressible fluid. Turbulence is modeled using the RNG k-ε turbulence model. The detailed GRI 3.0 chemical kinetic network with 325 reactions and 53 species is used to model the non-premixed combustion in the steam cracking firebox. To achieve reasonable computational effort, it is impossible to model every ordinary differential equation for all components in every cell at each iteration, so this mechanism is implemented in a tabulated way using non-adiabatic steady diffusion flamelet modeling.[4] Thermal NOx formation is modeled using the Zeldovich mechanism[5], using separate conservation equations calculated as a post-processing step, as implemented within FLUENT. The discrete ordinates model is used to solve radiative heat transfer.[6]

The geometry of the reactor coil is that of a multi-pass coil with 2 inlet tubes per outlet tube. A total of 16 floor burners and 32 sidewall burners provide the heat to the reactor.

Results and discussion

Effects of air distribution to floor burners

Simulations using different primary/secondary air ratios on the floor burners are performed to check their effect on furnace performance. The impact on flame shape, tube metal temperatures, run length and NOx formation is examined. To compare, two primary/secondary air ratios will be used: 25/75 and 75/25. The primary/secondary air ratio of the sidewall burners is kept at 50/50.

An evaluation of the streamlines in the furnace indicates that a large recirculation zone appears in the middle of the furnace when using more secondary air (see Figure 1). This affects the interaction between the floor and sidewall burners, and consequently the incident radiative heat flux (IRHF) to the coils. The simulations indicate that the outlet passes in the 25/75 case have a more uniform IRHF over the furnace height, with the top and bottom of the furnace having a higher IRHF compared to the 75/25 case. This results in a lower maximum TMT in the 25/75 case, which in turn results in significantly reduced coke formation and hence a 66% increase in run length compared to the 75/25 case.

When looking at the simulated NOx concentrations at the exit of the firebox, a significant decrease is found when decreasing the amount of primary air. The 75/25 primary/secondary air ratio results in a calculated NOx amount of 314ppm, compared to 205ppm for the 25/75 primary/secondary air ratio, which is a 53% increase by adjusting the air registers to the burners.

Effects of air distribution to sidewall burners

To check the effect of air distribution towards the sidewall burners, two different primary/secondary air flowrate ratios have been tested: 100/0 and 50/50. For the floor burners a primary/secondary air ratio of 25/75 is used because this was the best case according to the above explanation.

The 100/0 case results in spread-out flames, covering the whole sidewall area due to the high velocity of the premixed fuel/air mixture exiting the burner head. In the 50/50 case on the other hand more compact flames are observed. The simulations also show a significant impact of the floor burner flow pattern and the corresponding recirculation zone in the middle of the furnace. The corresponding different IRHF profile obtained by altering the air flowrates to the sidewall burners again results in a different coking rate over the reactor and therefore a different run length. The effect is however smaller than the effect of the floor burner air distribution: a 13% longer run length is observed for the 100/0 case compared to the 50/50 case.

There is a decrease in NOx formation with increasing primary air, i.e. 205ppm for the 50/50 case and 143ppm for the 100/0 case. This is caused by the increased velocity of the fuel-air mixture exiting the sidewall burner head. An increased velocity results in a more spread-out flame. Although the peak temperature is higher in the 100/0 case, the residence time in this high temperature region is so low that a lower amount of NOx is formed. Therefore the shift to more primary air results in a NOx reduction for the sidewall burners.

Conclusions

In this contribution, highly detailed 3D CFD simulations are performed of an industrial steam cracking furnace. The effect of the air distribution is investigated by manipulation of the primary to secondary air ratio to the floor burners and sidewall burners separately. It was shown that this manipulation results in a significant impact on the performance of the firebox. This is not only applicable to steam cracking, but can be extended to steam reformers, EDC cracking furnaces, fired heaters in refineries etc.

References

1. Baukal, C. E. J., The John Zink Hamworthy Combustion Handbook. CRC press: 2012.
2. Sullivan, B. K., Ethylene Cracking Heater Troubleshooting. In AIChE Spring National Meeting 2015, Austin, Texas, 2015.
3. Vinayagam, K., Minimizing Flame Impingements In Fired Heaters. Chemical Engineering 2007, 114 (5), 70-73
4. Pitsch, H.; Barths, H.; Peters, N. Three-dimensional modeling of NO x and soot formation in DI-diesel engines using detailed chemistry based on the interactive flamelet approach. SAE transactions 1996, 2010.
5. Zeldvich, Y. B., The oxidation of nitrogen in combustion and explosions. J. Acta Physicochimica 1946, 21, 577.
6. Carlson, B. G.; Lathrop, K. D. Transport theory: the method of discrete ordinates; Los Alamos Scientific Laboratory of the University of California, 1965.

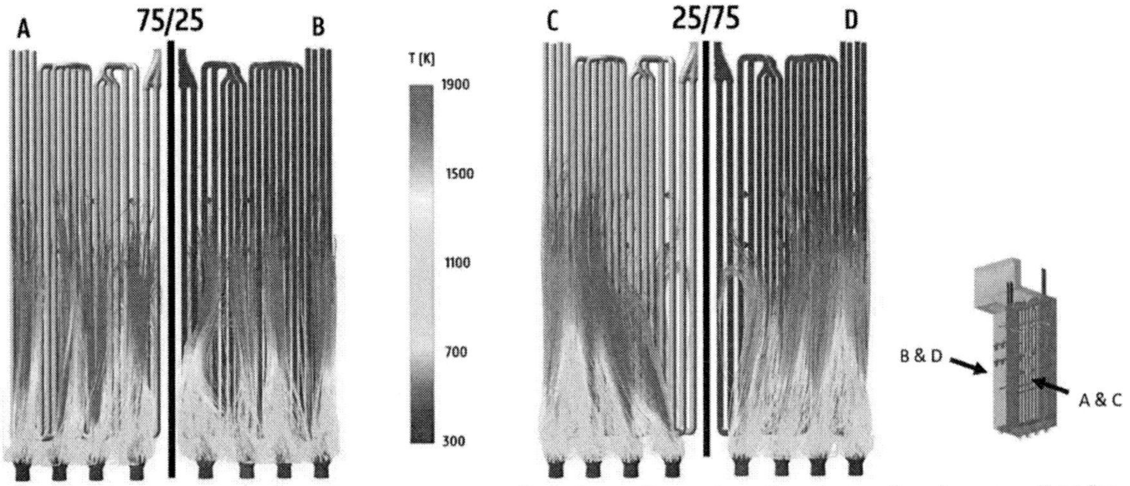

Figure 1: Air streamlines colored by temperature using primary/secondary air ratios on floor burners of: 75/25 (A&B) and 25/75 (C&D). A&B and C&D correspond to both sides of the simulated furnace, on either side of the reactor coil as indicated on the bottom right. A large recirculation zone appears in the middle of the furnace when using more secondary air.

622182 Methane Dry Reforming (MDR) with the *Ni-Ugso* Catalyst: Kg-Lab Scale Tests at Industrial H_2 Production Conditions

Abstract:

This work belongs to a larger endeavor aimed at studying the performance and the stability of a patent pending hydrocarbons reforming spinellized nickel catalyst prepared from an ilmenite metallurgical residue. Recently published results from experiments carried out with this catalyst exhibited high hydrogen production selectivity during methane steam, dry and autothermal (CO_2 and O_2) reforming and stability over time-on-stream (TOS).

At the lab-sale barometric conditions and at relatively lower severity than the prevailing industrial regimes, that is T=810 °C, space velocity of 3400 mL/h.g_{cat} and molar CO_2/CH_4=1.2, Ni-UGSO gives a stable methane conversion of 92% over 168h TOS. In this DRM reaction optimization study the best performances are obtained with 10-13wt% Ni load.

This work focuses on scaling up the process by a factor of 1000 (from g-lab to kg-lab scale). Various reforming regimes are tested but this work presents the endeavor on DRM, which aims at proving that the new catalyst (Ni-UGSO) can contribute in rendering DRM a commercially viable alternative to the prevailing H_2 production processes from natural gas steam reforming. The main challenges are to prove that:

- This much cheaper catalytic formulation (1) can be produced in forms usually used commercially by means of standard protocols and (2) has at least the same efficiency, robustness and life cycle characteristics with the best commercially available reforming catalysts;
- The DRM proposed is at least as effective as steam reforming.

This work presents (a) the scale-up calculations and criteria used based on the lab-scale results; (b) the production of the catalyst in pellets as well as their critical mechanical properties for its eventual use in industrial fixed bed reactors; (c) the kg-lab set-up and operation protocol and (d) the results at kg-lab scale towards answering the above challenges.

Full mass balances, statistical validation of the results and full analyses of the products and the catalysts (fresh and used) are included. BET, SEM-EDXS and XRD analyses are combined to provide data regarding the behavior and aging of the catalyst at conditions typically used in industry for accelerated aging.

626796 Modeling and Prediction of Asphaltene Precipitation for Two Crude Oil Samples from Kuwaiti Fields Along Production Wells

Abstract:

Two samples of crude oils were collected from the gathering centers for analysis and asphaltene stability tests. Properties such as density, viscosity, asphaltene and SARA analysis were determined in laboratory. One of the crude oil sample was heavy with API Gravity of 20 and asphaltene content of 9.7% while the other oil sample had API gravity of 27.7 and asphaltene content of 1.8%. Asphaltene stability tests by 5 different methods showed both oils were unstable with respect to asphaltene precipitation. A thermodynamic model based on SAFT equation of state was programmed using excel to demonstrate phase behavior of these asphaltenic oils. In addition, a model was developed to predict temperature and pressure profiles along wellbore which was favorably compared with results obtained from professional simulators such as PIPESIM and OLGA. In this modeling heat and mass transfer equations are solved numerically. The temperature drop is calculated from heat loss from the fluid and pressure drop is estimated from friction and hydrostatic pressure drops with the help of Duns and Ros method. By combination of predicted PT profile with asphaltene phase diagram we can show the exact region along the wellbore where aphaltene phase separation occurs.

Optimal modification on acidity of Y zeolite-based catalysts for high alkylation stability

Chengxi Zhang*, Yongxiang Li, Shunli Zhou, Yibin Luo and Xingtian Shu

Research Institute of Petroleum Processing, Haidian District, Beijing, 100083, China

Zeolites catalyzed isobutane-butene alkylation has attracted much research in oil refining and catalysis fields in recent years. Development of high stability zeolite-based catalysts is the key to the industrialization of this technology. Based upon our results, catalytic life is closely related to the competition between hydride transfer and oligomerization. Optical modification on the catalysts' acidity is the key to obtain high alkylation stability, however the major challenge lies in the fact that the acidity would be greatly influenced upon modification. Herein, a new idea on the modification of the whole catalyst is proposed to tailor the acidity and pore structure of catalyst through the zeolites-alumina interaction.

The whole catalyst modification upon zeolites-alumina interaction was applied, the effect of Al_2O_3 binders' particle size and SiO_2 doping on acidity and alkylation performance of Zeolite-Y based catalysts were studied. The catalysts were characterized by N_2 adsorption/desorption, X-ray powder diffraction (XRD), pyridine adsorbed FT-IR and MAS NMR. The interaction between Y and Al_2O_3 was enhanced with the decreasing Al_2O_3 particle size. The unit cell parameters increase with the decreasing Al_2O_3 particle size. At the same time, the acidity of catalysts increases with decreasing Al_2O_3 particle size. The Al_2O_3 phase's transition temperature was increased with SiO_2 doping, which enhance the interaction between Al_2O_3 binders and zeolite. The shrinkage of the unit cell was inhibited. The acid amount of the catalysts was far more improved comparing with undoped samples, and the optimal doping amount was 12%. The interaction mechanism between Zeolite Y and Al_2O_3 binder was proposed: (i) the interaction occurred only at phase α-$Al_2O_3 \cdot H_2O$ during calcination, $Al(OH)_3$ from Al_2O_3 could immigrate into zeolite, react with Si-OH and form new Brönsted acid sites. (ii) Particle dispersion was enhanced by decreasing the particle size of α-$Al_2O_3 \cdot H_2O$, which could strengthen the interaction between zeolite and Al_2O_3 binder. (iii) The phase transformation temperature from α-$Al_2O_3 \cdot H_2O$ to γ-Al_2O_3 was improved by the doping of oxide, which could increase the interaction time, and enhance $Al(OH)_3$ immigration and the interaction between zeolites and Al_2O_3 binder. The alkylation lifetime of the modified catalyst upon zeolites-alumina interaction was 38 h longer than that of the original sample, and the C_8 selectivity was 4 percentage point higher. This study will also provide a scientific basis for the regulation of acidity of other solid acid catalysts.

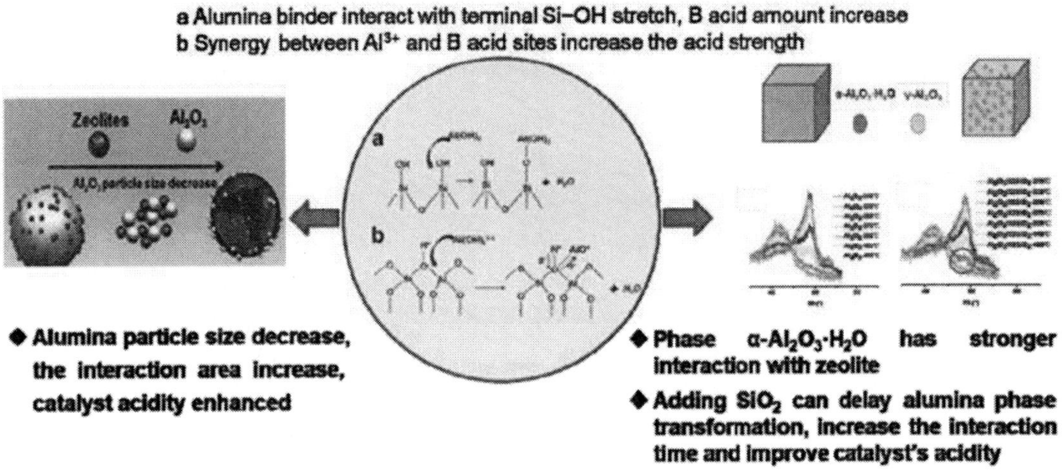

Figure 1. zeolite-alumina interaction scheme

620496 An Experimental and Modeling Study on Low- and High-Temperature Pyrolysis and Oxidation of Oxymethylene Ethers As a Sustainable Alternative Fuel

Abstract:

Introduction

The present research focuses on the combined modeling and experimental work on the pyrolysis and oxidation of oxymethylene ethers (OMEs). OMEs are a family of molecules with alternating carbon and oxygen atoms in the backbone saturated with hydrogen atoms, represented with structural formula $CH_3O(CH_2O)_nCH_3$ (further indicated as OME-*n*). These molecules have interesting properties for applications as alternative fuel. OMEs are categorized as e-fuels since they can be produced in a carbon-neutral manner via carbon capture and utilization (CCU) starting from captured CO_2 and renewable energy. In addition, blending them with conventional diesel reduces toxic soot emissions, due to the absence of carbon-carbon bonds, while still being compatible with the current generation of diesel engines. OMEs could as such contribute to the development of a more sustainable transport sector within the near future, and a circular carbon economy in general. However, before being widely applicable, it is important to fully understand the pyrolysis and oxidation chemistry of these compounds and their interactions with hydrocarbons. Therefore, a new microkinetic model is developed based on first principles that describes the low- and high-temperature decomposition pathways of oxymethylene ethers up to OME-5. This model enables predictive simulations for combustion applications.

Methodology

During the last decade, great progress has been made in the construction of reliable kinetic models for numerous technologically important radical chemistry processes. The resulting models typically contain hundreds of species, and several thousands of associated reactions. The generation of microkinetic models by hand is thereby a tedious, error-prone and often incomplete process. To facilitate this process, automatic kinetic model generation codes such as Genesys at the Laboratory for Chemical Technology have been developed to systematically generate kinetic models [1]. For this research, the thermodynamic and kinetic databases of Genesys are extended with more relevant data for C-O-C and O-C-O bonds, to enable construction of microkinetic models for oxymethylene ethers based on high-level quantum chemical data.

A prerequisite for the generation of detailed kinetic models is the availability of accurate thermodynamic and kinetic data for species and reactions, respectively. Ideally, these parameters are available from experiments or high-level quantum chemical calculations. Since these methods are expensive and time-consuming, Genesys instead often relies on approximation methods such as group additivity theory and rate rules. In this work, thermodynamic and kinetic parameters are obtained from quantum chemical calculations at the CBS-QB3 level of theory for important reaction pathways of both low- and high-temperature oxidation of OMEs. The lowest energy conformer is searched extensively by in-house developed tools. Non-harmonic normal modes are accounted for by performing one-dimensional hindered-rotor calculations. The generated data – thermodynamic properties for around 700 species and modified Arrhenius parameters for around 120 reactions – is extrapolated to become valid for longer chain OMEs by regression of new values for Benson's group additivity and rate rules.

Results

Within Genesys, the possible reactions are generally defined in terms of reaction families. Reaction families from earlier studies on smaller compounds with C-O-C bonds (i.e., dimethyl ether and dimethoxy methane/OME-1 [2]) with Genesys are extended with new unimolecular decomposition pathways. The latter were identified as depicted on the potential energy surface for OME-2 in Figure 1. Roaming reactions are observed with the formation of products with a hydroxyl functionality and carbenes. These carbenes (except for methylene) can decompose with the formation of radicals. In addition, three other concerted reactions are found forming a carbonyl functionality in addition to a smaller ether. The reaction from OME-2 to OME-1 and formaldehyde can proceed via three different transition states since each carbon-oxygen bond can be eliminated. For the thermal decomposition of OMEs, these reactions are dominant pathways, because several of them have an electronic barrier that is significantly lower than the lowest bond dissociation energy, i.e., 348 kJ mol^{-1}. This clarifies the high reactivity already at low temperatures and indicates that radical chemistry is less dominant at these temperatures. Eventually, the outcome is a model containing the chemistry for OME-1 up to OME-5. To include detailed chemistry of smaller (oxygenated) hydrocarbons in the final model, the Genesys model is merged with the AramcoMech 1.3 base model [3]. In case of conflicting thermodynamic or kinetic properties, the input of the AramcoMech model is kept to maintain the integrity of this experimentally fitted model.

Various pyrolysis and oxidation experiments have been performed to validate both the primary and secondary chemistry of the generated model. Low-temperature pyrolysis experiments are performed for OME-2, OME-3, OME-4 and OME-5 in the micropyrolysis set-up at the Laboratory for Chemical Technology (LCT). Small amounts of fuel are vaporized, diluted, and brought in a tubular reactor where the decomposition products are measured over the broad temperature range of 373 K to 873 K at a pressure of 3.4 bara. The main products are shorter ethers, formaldehyde, methane, methanol and at higher temperatures hydrogen, carbon monoxide and carbon dioxide due to secondary decomposition reactions. Preliminary results indicate significant decomposition of OME-2 starting at around 450 K, as depicted in Figure 2. Similarly, high-temperature pyrolysis experiments are performed for OME-2 and OME-3 in the bench-scale steam cracker setup at LCT over a range of temperatures 723 K to 1073 K at 1.5 bara. A similar product spectrum is identified accompanied by minor amounts of OME-derived species with carbonyl and hydroxyl functionalities, e.g., formic acid, methyl formate, methoxymethanol and methoxymethyl formate. The number of products with carbon-carbon bonds detected and the associated quantities are low.

Ignition delay times have been measured via rapid compression experiments at 5 bara for and for OME-2 and OME-3 to validate the low-temperature oxidation section of the model. Samples were taken during compression to identify the most important reactants and products. These include, in the case of OME-3, methyl formate, methoxymethyl formate and methoxymethoxymethyl formate. From quantum chemical calculations, it follows that the formation of cyclic ethers plays an important role which is confirmed qualitatively from the measured products. Cyclic ethers derived from OMEs tend to decompose quickly by formation of carbonyl functionalities, similar to a reverse Diels-Alder reaction.

Finally, flat flame burner experiments have been performed with OME-3 for validation of the high-temperature combustion chemistry. The flame experiments are performed at 0.0533 bara and with a fuel composition of 20 mol% OME_3 and 80 mol% CH_4. Concentration profiles are constructed as function of height above burner. The main compounds detected are OME-3, formaldehyde, methanol, hydrogen, carbon dioxide and carbon monoxide. While ethane, ethylene, dimethyl ether, methyl formate, dimethoxy methane and methoxymethyl formate are observed in minor amounts. Preliminary simulation results indicate good agreement between the model and experimental observations for major and minor species without fitting of model parameters.

Conclusions

The thermodynamic and kinetic databases of Genesys are extended to enable the construction of microkinetic models for oxymethylene ethers. A microkinetic model has been constructed with Genesys for pyrolysis and both low- and high-temperature oxidation of oxymethylene ethers up to OME-5 based on first principles. The model is validated with several series of experiments including pyrolysis in tubular reactors, ignition delay time measurements in a rapid compression machine and flat flame burner experiments.

Acknowledgments

This work was established in a joint call to strengthen collaboration between I-SITE Université Lille Nord-Europe (members and national partners) and Ghent University. Kevin De Ras and Ruben Van de Vijver acknowledge the Fund for Scientific Research Flanders (FWO) respectively via doctoral fellowship grant 3F018119 and post-doctoral fellowship grant 3E013419. The research leading to these results has received funding from the European Research Council under the European Union's Horizon 2020 research and innovation programme / ERC grant agreement n◦ 818607.

References

[1] N. M. Vandewiele, K. M. Van Geem, M.-F. Reyniers, and G. B. Marin, "Genesys: Kinetic model construction using chemo-informatics," *Chemical Engineering Journal,* vol. 207-208, pp. 526-538, 2012/10/01/ 2012.

[2] F. H. Vermeire, H.-H. Carstensen, O. Herbinet, F. Battin-Leclerc, G. B. Marin, and K. M. Van Geem, "Experimental and modeling study of the pyrolysis and combustion of dimethoxymethane," *Combustion and Flame,* vol. 190, pp. 270-283, 2018/04/01/ 2018.

[3] W. K. Metcalfe, S. M. Burke, S. S. Ahmed, and H. J. Curran, "A Hierarchical and Comparative Kinetic Modeling Study of C1 − C2 Hydrocarbon and Oxygenated Fuels," *International Journal of Chemical Kinetics,* vol. 45, no. 10, pp. 638-675, 2013.

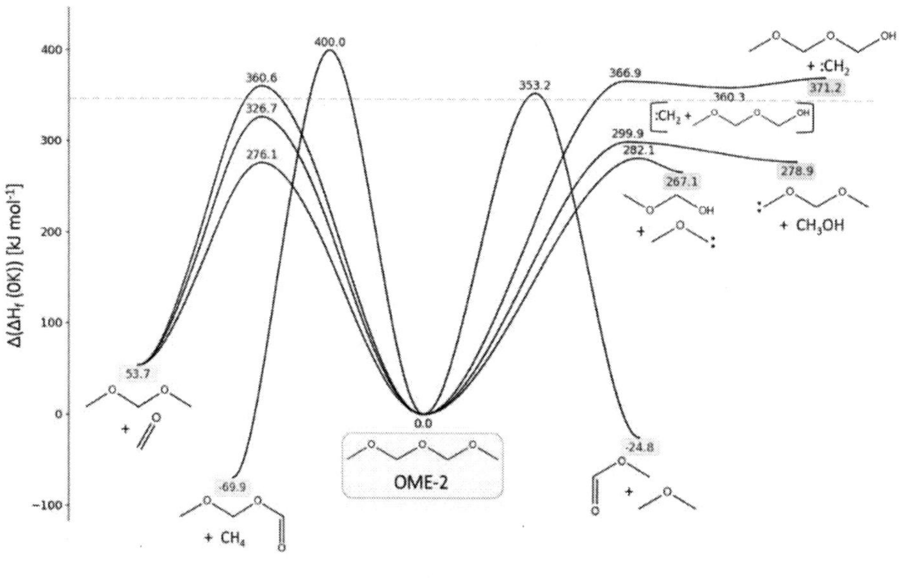

Figure 1: Potential energy surface at 0 K for OME-2 at the CBS-QB3 level of theory. Energies are relative to OME-2. The dashed line indicates the lowest bond dissociation energy of 348 kJ/mol. Products between brackets indicate the formation of a van der Waals-well. Energies within a green box correspond with roaming reactions, energies in a yellow box correspond with carbonyl functionality formation.

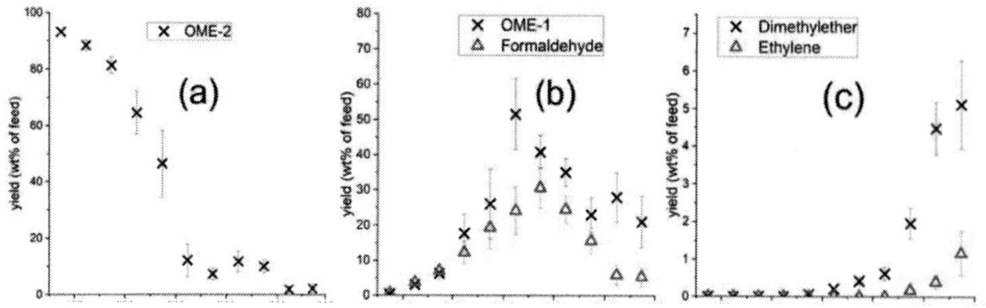

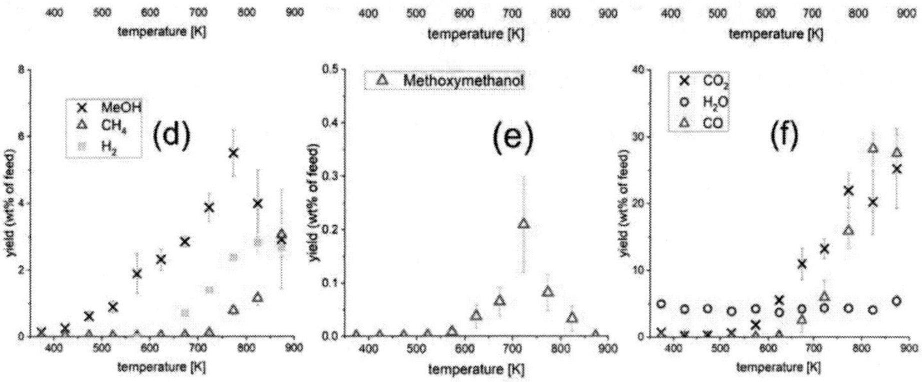

Figure 2: Preliminary results of the decomposition products of OME-2 obtained from pyrolysis of OME-2 diluted with helium in the temperature range 373 to 873 K at a pressure of 3.4 bara.

628610 Life-Cycle and Techno-Economic Analysis of Fischer-Tropsch Fuel Production Using Waste CO_2

Abstract:

Electro-fuel (E-fuel) has received increasing interest due to its potential to reduce dependence of on fossil fuels and to help mitigate climate change. Among commonly researched E-fuels, Fischer–Tropsch (FT) fuel is most compatible with conventional transportation fuels widely used for heavy-duty vehicles and non-road transportation applications.

This study modeled the FT fuel synthesis process using Aspen Plus. The life cycle assessment (LCA) and techno-economic analysis (TEA) have been conducted using GREET 2020 and H2A economic analysis model. The U.S. national FT fuel production capacity and carbon reduction potential has been evaluated by considering CO_2 captured from six industrial applications (ethanol, ammonia, natural gas [NG] processing, hydrogen, cement, and iron/steel production plants), and two fossil power generation technologies (coal and NG).

Our results show that FT fuels from various plant designs using renewable or nuclear power have near zero greenhouse gas emissions. However, when the H_2 price is at $2.0/kg (2020 DOE target), the MFSP of the FT fuel is 74-90% higher than the pre-tax petroleum diesel price of $3.1/gal in 2050. A total of 1,600 million metric ton per year of waste CO_2 can be captured and converted into 85 billion gallons of FT fuels in U.S. The FT fuels can potentially substitute for 36% of the total U.S. petroleum fuels used in the transportation sector. The total CO_2 reduction potential associated with FT fuels production is 935 MMT/year when replacing corresponding fossil fuels.

628218 Dewatering/Dehydration of Bioethanol Using Ultrasound

Abstract:

Bioethanol production from renewable sources plays an important role in the US effort to facilitate sustainability, mitigate environmental impact, and enhance energy independence. Current technologies used for bioethanol purification include traditional distillation, extractive distillation, azeotropic distillation, and pressure swing distillation, among others. However, each of these has significant limitations, such as high energy consumption and low energy efficiency, inefficient separation, requirements for other chemicals, operational complexity, limited lifetime of materials, etc.

When an ultrasonic wave generated from a piezoelectric actuator is applied to an ethanol-water solution, fine droplets are formed due to capillary wave instability development on the ethanol-water interface and acoustic cavitation. The ethanol concentration in the ultrasound-generated fine droplets is found to be greater than that in the bulk solution, apparently because of an ethanol-enriched shell.

In this study, a batch ultrasonic separation prototype was first developed to determine the feasibility of ultrasound-driven ethanol purification, and the effects of operational parameters, such as liquid temperature, liquid thickness, location of the carrier gas inlet/outlet, and carrier gas flow rate on the enrichment of ethanol in collected mist. Following that, three types of continuous-flow ultrasonic separation prototypes, including a 10-transducer separation unit, were developed to demonstrate the feasibility of replacing the current distillation process. Results from the batch unit show that liquid temperature and carrier gas flow rate significantly affect ethanol concentration in collected mist and the collection rate. A simple process analysis shows that a two-stage continuous-flow ultrasonic separation unit can potentially replace one distillation column in the traditional three-column ethanol purification process, and a five-stage continuous-flow unit could replace two distillation columns in the current process. The ten-transducer separation unit was shown to significantly reduce the effect of carrier flow rate on ethanol enrichment. A preliminary techno-economic analysis shows that this technology has the potential to reduce energy consumption by 30% compared to the current process. The application of ultrasound in ethanol purification may provide an alternative to the current ethanol purification process, operating at atmospheric pressure and at or slightly above ambient temperature.

628146 Rational Design of Aqueous Ionic Liquids for Biogas Upgrading Using a Molecular-Based Equation of State

Abstract:

Over the past decades, the potentiality of task-specific solvents, such as ionic liquids (ILs) and deep eutectic solvents (DESs), for carbon dioxide (CO_2) capture and separation, with the potential of replacing conventional aqueous amines [1]. Their attractiveness for biogas upgrading is due to their high CO_2 uptake, and easily tunable and unique thermophysical properties such as low volatility and high thermal and chemical stability compared to aqueous amines [2,3]. Notwithstanding, a primary issue with the application of ILs for CO_2 separation is their high viscosity, which has an adverse effect on the mass transfer between CO_2 and the solvent, directly affecting the capital and operating costs of the process [4]. Mixtures of ILs with other physical solvents such as water or alcohols have the potential to effectively reduce the viscosity of these solvents, however, this would be on the expense of reducing their CO_2 capturing ability [5]. This would require fine-tuning the ratio of both solvents, taking into account the inherent trade-off between CO_2 uptake, viscosity, and other criteria such as absorption selectivity, and regeneration energy requirements. Accomplishing such a task using experimental techniques, considering the large number of available ILs and their possible hybrid mixtures with physical solvents is time and cost intensive. However, with the current advancements in computational power and molecular modeling approaches, the accomplishment of such as task is made easier, allowing the systematic evaluation of the effect of the addition of a physical solvent on the performance of ILs for a particular application.

In this contribution, we demonstrate the application of a robust molecular-based equation of state (EoS), namely, soft-SAFT EoS [6,7], towards the rational design of aqueous ionic liquids for biogas upgrading. The thermodynamic modeling of pure fluids within the framework of soft-SAFT is done in a coarse-grain manner by representing fluids by a set of molecular parameters characterizing key structural and energetic features. The examined ILs, $[C_4mim][BF_4]$, $[C_4mim][PF_6]$, $[C_2mim][NTf_2]$, and $[C_2mim][CF_3SO_3]$, were modelled as associating chainlike fluids [8] , CO_2 was modelled as a chainlike molecule with explicit consideration of its quadrupolar nature [9], hydrogen sulfide (H_2S) was modelled as an associating fluid, methane (CH_4) was modelled as a non-associating, non-polar spherical segment, while water was modelled as an associating spherical segment [10,11]. The thermodynamic modeling was done in a systematic manner initially through validating the performance of the thermodynamic model with available experimental data [5,12,13]. This examination included solubility of CO_2, H_2S, and CH_4 in pure ILs and pure water, phase equilibria and viscosity of binary mixtures of ILs and water, solubility of the aforementioned gases in aqueous ILs. The assured reliability of the thermodynamic model permitted the predictive application of the model in a systematic examination of the effect of water content in aqueous ILs on their performance for biogas upgrading. This was inclusive of a wide range or properties such as solvent viscosity, solubility of CO_2, CH_4, and H_2S, along with absorption selectivity, and lastly enthalpy of CO_2 absorption.

With these criteria, the water content in aqueous ILs was optimized to ensure the maximum gain from the inherent trade-off with the addition of water to ILs in terms of reduced CO_2 solubility, and reduced solvent viscosity.

This framework with a molecular-based EoS at its center attests to the efficacy of employing molecular modeling approaches to the rational design of solvents for CO_2 capture and separation, that assists in a more targeted deployment of experimental efforts on promising solvents.

This work is funded by Khalifa University of Science and Technology (RC2-2019-007). Computational resources from the Research and Innovation Center on CO_2 and H_2 (RICH Center) are gratefully acknowledged.

References

[1] S. Sarmad, J.-P. Mikkola, X. Ji, CO2 capture with Ionic liquids (ILs) and Deep Eutectic Solvents (DESs): a new generation of sorbents, ChemSusChem. 10 (2017) 324–352. https://doi.org/10.1002/cssc.201600987.

[2] L.F. Vega, O. Vilaseca, F. Llovell, J.S. Andreu, Modeling ionic liquids and the solubility of gases in them: Recent advances and perspectives, Fluid Phase Equilib. 294 (2010) 15–30. https://doi.org/10.1016/j.fluid.2010.02.006.

[3] I.I.I. Alkhatib, D. Bahamon, F. Llovell, M.R.M. Abu-Zahra, L.F. Vega, Perspectives and guidelines on thermodynamic modelling of deep eutectic solvents, J. Mol. Liq. 298 (2020) 112183. https://doi.org/10.1016/j.molliq.2019.112183.

[4] M.T. Mota-Martinez, P. Brandl, J.P. Hallett, N. Mac Dowell, Challenges and opportunities for the utilisation of ionic liquids as solvents for CO2 capture, Mol. Syst. Des. Eng. 3 (2018) 560–571. https://doi.org/10.1039/c8me00009c.

[5] M. Aghaie, N. Rezaei, S. Zendehboudi, Assessment of carbon dioxide solubility in ionic liquid/toluene/water systems by extended PR and PC-SAFT EOSs: Carbon capture implication, J. Mol. Liq. 275 (2019) 323–337. https://doi.org/10.1016/j.molliq.2018.11.038.

[6] F.J. Blas, L.F. Vega, Thermodynamic behaviour of homonuclear and heteronuclear Lennard-Jones chains with association sites from simulation and theory, Mol. Phys. 92 (1997) 135–150. https://doi.org/10.1080/002689797170707.

[7] F.J. Blas, L.F. Vega, Prediction of Binary and Ternary Diagrams Using the Statistical Associating Fluid Theory (SAFT) Equation of State, Ind. Eng. Chem. Res. 37 (1998) 660–674. https://doi.org/10.1021/ie970449+.

[8] I.I.I. Alkhatib, M.L. Ferreira, C.G. Alba, D. Bahamon, F. Llovell, A.B. Pereiro, J.M.M. Araújo, M.R.M. Abu-Zahra, L.F. Vega, Screening of Ionic Liquids and Deep Eutectic Solvents for Physical CO2Absorption by Soft-SAFT Using Key Performance Indicators, J. Chem. Eng. Data. 65 (2020) 5844–5861. https://doi.org/https://doi.org/10.1021/acs.jced.0c00750.

[9] I.I.I. Alkhatib, F. Llovell, L.F. Vega, Assessing the effect of impurities on the thermophysical properties of methane-based energy systems using polar soft-SAFT, Fluid Phase Equilib. 527 (2021) 112841. https://doi.org/10.1016/j.fluid.2020.112841.

[10] L.F. Vega, F. Llovell, F.J. Blas, Capturing the solubility minima of n-Alkanes in water by Soft-SAFT, J. Phys. Chem. B. 113 (2009) 7621–7630. https://doi.org/10.1021/jp9018876.

[11] L.F. Vega, F. Llovell, Review and new insights into the application of molecular-based equations

of state to water and aqueous solutions, Fluid Phase Equilib. 416 (2016) 150–173. https://doi.org/10.1016/j.fluid.2016.01.024.

[12] M. Watanabe, D. Kodama, T. Makino, M. Kanakubo, CO2 absorption properties of imidazolium based ionic liquids using a magnetic suspension balance, Fluid Phase Equilib. 420 (2016) 44–49. https://doi.org/10.1016/j.fluid.2015.12.055.

[13] A. Yokozeki, M.B. Shiflett, C.P. Junk, L.M. Grieco, T. Foo, Physical and chemical absorptions of carbon dioxide in room-temperature ionic liquids, J. Phys. Chem. B. 112 (2008) 16654–16663. https://doi.org/10.1021/jp805784u.

AUTHOR INDEX